Using Visual Evidence

ONE WEEK LOAN

Using Visual Evidence

Edited by Richard Howells and Robert W. Matson

Open University Press

Open University Press
McGraw-Hill Education
McGraw-Hill House
Shoppenhangers Road
Maidenhead
Berkshire
England
SL6 2QL

email: enquiries@openup.co.uk
world wide web: www.openup.co.uk

and Two Penn Plaza, New York, NY 10121–2289, USA

First published 2009
Copyright © Richard Howells and Robert Matson 2009

A catalogue record of this book is available from the British Library

ISBN–13: 9780335228645 (pb)   9780335228638 (hb)
ISBN–10: 033522864X (pb)   0335228631 (hb)

Library of Congress Cataloging-in-Publication Data
CIP data applied for

Typeset by YHT Ltd, London
Printed in the UK by Bell and Bain Ltd, Glasgow

The McGraw·Hill Companies

For Lisa and Sarah

Contents

Figures

Contributors

Stuart Allan is Professor of Journalism in the Media School at Bournemouth University. He is the author of *News Culture* (1999, 2004), *Media, Risk and Science* (2002) and *Online News: Journalism and the Internet* (2006). His edited collections include *Journalism After September 11* (2002, with B. Zelizer), *Reporting War: Journalism in Wartime* (2004, with B. Zelizer) and *Journalism: Critical Issues* (2005). He is a book series editor for the Open University Press and serves on the editorial boards of several journals.

Jacqueline Botterill is Assistant Professor in the Department of Communication, Popular Culture and Film at Brock University, Canada. She is the author of *Consumer Culture and Personal Finance* (2008) and co-author of *Dynamics of Advertising* (2000, with B. Richards and I. MacRury) and *Social Communication in Advertising* (3rd ed., 2005, with W. Leiss, S. Kline and S. Jhally).

Cynthia Carter is Senior Lecturer in the School of Journalism, Media and Cultural Studies at Cardiff University. Her research interests include children, news and citizenship; feminist media studies; and violence and the media. She is co-author of *Violence and the Media* (2003, with C.K. Weaver). Her edited books include *Critical Readings: Media and Gender* (2004, with L. Steiner) and *Critical Readings: Violence and the Media* (2006, with C.K. Weaver). She is founding co-editor of *Feminist Media Studies* and is an editorial board member of several journals.

Máire Messenger Davies is Professor of Media Studies, Director of the Centre for Media Research and Director of the Media Studies Research Institute at the University of Ulster, Northern Ireland. A former journalist, she earned her BA in English from Trinity College Dublin and a PhD in Psychology from the University of East London. Her books include *Television is Good for your Kids* (1989, 2001), *Fake, Fact and Fantasy: Children's Understanding of Television Reality* (1997) and *'Dear BBC': Children, Television-storytelling and the Public Sphere* (2001).

Nicholas Hiley is a social historian specializing in the history of censorship and the mass media, who has worked and taught at the University of Cambridge, King's College London and the University of Kent. He is currently Head of the British Cartoon Archive, based at the University of Kent, which holds the national collection of 120,000 original cartoons of political and social comment from British newspapers and magazines. Its online catalogue of over 130,000 cartoons from the eighteenth to the twenty-first centuries is freely available at www.cartoons.ac.uk.

Richard Howells is Reader in Cultural and Creative Industries at King's College London, where he is also Director of the Centre for Cultural, Media and Creative Industries Research. He has an international background in sociology, visual and popular culture. In 2004 he was Distinguished Visiting Professor at the Centre for the Arts in Society at

Carnegie Mellon University. His publications include *The Myth of the Titanic* (1999) and *Visual Culture* (2003).

Stephen Kline is Director of the Media Analysis Lab, School of Communication at Simon Fraser University, British Columbia, Canada. Widely published in the fields of advertising, marketing and children's culture, he is the author of *Out of the Garden: Toys, TV and Children's Culture in the Age of Marketing* (1995) and *Fast Food, Sluggish Kids* (2008) and co-author of *Social Communication in Advertising* (2005).

Luke McKernan is Curator, Moving Image at The British Library in London. His publications include *Topical Budget: The Great British News Film* (1992) and *Yesterday's News: The British Cinema Newsreel Reader* (1992).

Robert W. Matson is Professor of History at the University of Pittsburgh at Johnstown, where he received the President's Medal for Excellence in teaching in 2002. He is the author of *William Mulholland, A Forgotten Forefather* (1976) and *Neutrality and Navicerts: Britain, the United States and Economic Warfare, 1939–40* (1994). As senior associate editor of the journal *Film & History*, he edited a series of special numbers on Film and World War II (1997) and has published numerous articles, columns and reviews.

David Morgan is Professor of Religion at Duke University, Durham, North Carolina. The author of *Visual Piety* (1998), *Protestants and Pictures* (1999), *The Sacred Gaze* (2005) and *The Lure of Images* (2007). He has also edited and co-edited several volumes, including *Key Words in Religion, Media, and Culture* (2008) and *Re-Enchantment* (2008, with James Elkins). He is co-founder and co-editor of the journal *Material Religion* and co-editor of a book series at Routledge entitled 'Religion, Media, Culture'.

Douglas R. Nickel is the Andrew V. Rosenthal Professor of Modern Art at Brown University, Rhode Island. The former director of the Center for Creative Photography at the University of Arizona and previously a curator at the San Francisco Museum of Modern Art, his exhibitions include Carleton Watkins: The Art of Perception; Dreaming in Pictures: The Photography of Lewis Carroll; and Snapshots: The Photography of Everyday Life. A specialist in nineteenth-century photography, he is also the author of *Francis Frith in Egypt and Palestine: A Victorian Photographer Abroad* (2003).

Michael Paris is Professor of Modern History at the University of Central Lancashire. His research interests are in war and popular culture. His books include *Warrior Nation: Images of War in British Popular Culture* (2000), *Over the Top! The Great War and Juvenile Literature in Britain* (2004) and, as editor, *The First World War and Popular Culture* (1999) and *Repicturing the Second World War: Representations in Film and Television* (2007).

David Phillips is Senior Lecturer in Visual Theory at the University of East London. He has published on the history and theory of photography, contemporary art, and on sexuality and representation. He is currently completing a book on social documentary photography.

Pierre Sorlin is Professor of Sociology at the Université de la Sorbonne Nouvelle Paris III and works in the Audiovisual Department at the Instituto Parri in Bologna, Italy. His many publications include *Film in History: Restaging the Past* (1980), *European Cinema, European Societies* (1991), *Mass Media* (1994) and *Italian National Cinema, 1896-1996* (1996).

Catherine Whalen is Assistant Professor at the Bard Graduate Center for Studies in the Decorative Arts, Design and Culture in New York City. She received her PhD in American Studies from Yale University, where she was a co-ordinator of the Photographic Memory Workshop. A specialist in American visual and material culture studies and the history and theory of collecting, she has published articles in *Afterimage, Winterthur Portfolio, Nineteenth Century* and *Studies in the Decorative Arts*.

Introduction

By Richard Howells and Robert W. Matson

We could be forgiven for thinking that history is all about words. It is, after all, our capacity for language, more than any other trait, that elevates us as a species, and we have long been accustomed to looking back on the appearance of writing as the signal development that marks a society's emergency from the dimness of prehistoric times into the full light of history. Yet the importance we attach to that threshold reveals more about us than it does about the originators of writing. For them it was an incremental step forward, one that made the keeping of records more convenient and gradually created an entirely new scribal class. But the impact on *our ability to know* has been profound. Enquiring into what happened before the advent of the written word remains a very difficult, though tantalizing, endeavour. But through time, as words have successively been uttered, written and (eventually) committed to print, the concept of recorded history stimulated great efforts to conserve, compile and learn to interpret the written record. The discipline of history was created. But long before we began to write things down, we had learned to mark, record, mediate and represent our world in visual form. Even today, we are born with the capacity to see, but have later to learn how to read. Similarly, although children must be taught how to write, they know almost instinctively how to mark the world with fingers and paints. Indeed, there seems to be a universal need to record human experience in visual form, and this is as true of the cave paintings of Lascaux as it is of the multimedia visual culture of today.

The task of interpreting (and often discovering) the visual traces of the distant past has traditionally fallen to archaeologists and anthropologists, for whom the concept of visual evidence is not a new one. Preliterate cultures were both oral and visual, and as their spoken words have not survived them, the visual record is the only way preliterate societies can speak to us today. The visual, though, has remained important as evidence for the understanding of ancient and classical civilizations even after they learned to write things down. This is not only because of the physical decay of so many early written sources, but also because the visual culture of the distant past is so often taken more seriously by scholars than that of the recent past and the present. Indeed, the study of history could be seen as a sliding scale between the relative importance of the visual and the verbal in progressive favour of the written word. We can see this happening in western Europe through the Middle Ages and into the Renaissance, including the massive trigger point of the Gutenberg revolution of the fifteenth century. When we press forward into the era of mass literacy and mass publication (to say nothing of the public library, the archive, the photocopier, the personal computer and electronic mail), the written word seems almost supreme. As for the visual evidence, its authority – and certainly its utility – is typically regarded in inverse proportion to its modernity. This is a misapprehension that the current volume seeks to counter.

This historical inversion is one of the reasons we have chosen to concentrate on more recent forms of visual evidence in this volume. The value of the visual to an understanding of the distant past is not an argument that needs so vigorously to be made, but the same cannot be said for the visual culture of more recent times – and especially the present day. Here, historians have too often relegated the image to a supporting role: one of illustrating that which is explained more fully and more capably in the written text. Visual images might convey useful information to the untutored or illiterate, but the written and printed word must be the true sources of sophisticated understanding. This is increasingly ironic, given the enormous mass of visual evidence in the form of photographs and motion pictures that has been created during the past century, and even more so because the newest media – television, the Internet and ubiquitous portable digital devices of various kinds – place a new reliance on both orality and visuality alongside literacy. Our aim, then, is to redress the balance.

We must begin with the recognition that the relegation of images to mere illustration is a relatively recent phenomenon, certainly in terms of visual evidence. Photographers were among those most keenly aware of the situation. Walker Evans clearly found it a running battle when he wrote: 'For the thousandth time, let it be said that pictures which are really doing their work don't need words.'[1] Evans' *cri de cœur* was uttered in support of a photographic essay by fellow photographer Lee Friedlander in 1963, but the evidential value of Evans' own work had been similarly supported by writer James Agee in his introduction to *Let Us Now Praise Famous Men* from 1941. Here, Evans' photographs of American farmers in the Great Depression worked alongside Agee's prose – and are nowadays probably the more revered of the two. Yet even at the time, Agee was able to declare: 'The photographs are not illustrative. They, and the text, are co-equal, mutually independent, and fully collaborative.'[2] Another Evans – Harold Evans, a man vitally linked to the written word by his long service as editor first of the *Sunday Times* and later of the *Atlantic Monthly* – reaffirmed the point when he described the creation of his acclaimed 1998 work *The American Century*. Once the basic outline had been determined, the next step was the choice of photographs. Before he began to compose the written text dealing with a particular subject, Evans would position the selected photographs on a page and then determine if the space remaining provided room for the text he wanted to write. If not, the entire layout had to be changed. 'The technique of it was', he said, 'the constant tension between the number of words needed to tell an event and [the space] to give the photography the essential size and primacy it needed.'[3] Thus, let us be clear that it is far from our purpose to demote the written word in compensation for the imbalance that has existed. Rather, like Agee and Harold Evans, we advocate the use of the visual alongside the verbal evidence today. This is a call aimed not only at historians but also at anyone within the arts and human sciences who wishes to widen their evidential base and therefore their understanding of humankind, both ancient and modern. This is a book, then, that seeks both to advocate the usefulness of visual evidence to scholarship and, at the same time, to suggest ways in which this might in practice be accomplished.

Similarly, while this volume echoes Agee's sentiments on photography, it proceeds to apply them to all kinds of visual texts. We begin, obviously enough, with painting but soon strike less traditional ground with cartoons, comics and satirical prints, photography (authored, documentary and vernacular), film (newsreel, documentary and feature) and television (news, drama and advertising). This forms an approach that is generally pursued nowadays under the banner of visual culture. It is an approach that both expands the types of visual texts traditionally thought worthy of academic scrutiny, together with the methodologies considered appropriate for the job. David Morgan, who contributes the opening chapter to this volume, has previously defined visual culture as the study of the way in which images: 'help construct the worlds people live in and care about'.[4] Yet images do more than construct worlds – as if that were not enough! – they also reflect and represent them. With that in mind, Morgan is entirely right to lament the way in which 'many scholars ignore images in their research because their notion of evidence does not include pictures'.[5] It is all the more lamentable because, as he observed in another earlier work, such images need to be understood as 'historical traces' of the past worlds of they represent.[6]

History is an artificial means of analysing something which, in fact, no longer exists. History, however, leaves its imprint through evidence, and frequently that evidence is visual. Images are made by human beings and therefore have embedded in them much information about their makers. Thus, although images may be seen as primary sources for the study of what has been shown, all visual texts are primary sources for the study of their makers. This, we contend, is true of the earliest graphic arts through to photography film and television.

David Morgan is not, of course, the first or only person to advocate the usefulness of visual evidence to the scholarly research. Cultural historian Peter Burke, for example, has a long-established and distinguished record in this field. Yet the usefulness – the necessity, even – of visual literacy as a whole was promoted famously in Victorian England by John Ruskin, who in 1846 declared:

> the greatest thing a human soul ever does in this world is to *see* something, and tell what it *saw* in a plain way. Hundreds of people can talk for one who can think, but thousands can think for one who can see. To see clearly is poetry, prophecy, and religion – all in one. Therefore, finding the world of Literature more or less divided into Thinkers and Seers, I believe we shall find also that the Seers are wholly the greater race of the two.[7]

Again, we stress that in this volume we do not seek to create a sort of 'ethnic divide' between users and exponents of different types of evidence, but we do still wish to restore the visual to a prominent and valued status. To an extent, then, we seek to enable a rediscovery of the visual skills of the past in order to facilitate both the investigation of history and the understanding of the present.

The chapters which constitute this volume are media, rather than theoretically, based. More than that, they frequently work by way of case studies in order to illustrate specific examples of the authors' own uses of visual evidence. The reader is consequently invited to apply similar thinking to his or her own investigations. Our contributors come from a wide variety of disciplines and backgrounds and we have not sought to impose one voice upon them all. We consider this plurality to be a strength. All the contributions, however, are united by a commitment to the usefulness of the visual as evidence regardless of their particular perspectives.

They are also connected by another important underlying assumption which it is perhaps necessary to underline. The approach advocated in this volume does not necessitate the qualitative approval or even the artistic evaluation of the visual texts under scrutiny. In this way – and in contrast to established traditions of literary and artistic criticism – we do not necessarily seek to elevate the wide variety of frequently popular sources under examination on the strength of their previously unrecognized aesthetic merit. This is important in both anticipation and refutation of those critics of visual culture and 'cultural studies' who fear that we seek to dismantle the canon of the best that has been thought, said (and even painted) and replace it with a version of lesser but more popular quality. Certainly, some of our contributors may make passionate claims for the quality of the objects of their scholarly attention, but this is not an a priori requirement. Common to them all, however, is a recognition of the *value* rather than the necessary *quality* of their chosen texts to a greater understanding of history and society.

Similarly, we have sought to impose no methodological structures on our contributors and their chapters. It is for that reason that the reader may note varying degrees of emphasis on context and (to a lesser extent) audience reception in addition to a close reading of the visual texts themselves. Similarly, there are those who are more interested than others on authorial intent. Certainly, however, there is no preconditional assumption in this volume that the purpose of textual (or even contextual) analysis is to discover and to privilege the conscious intent of the original author as the real 'meaning' of the work. D.H. Lawrence famously identified 'opposing morals, the artist's and the tale's' in classic American literature before concluding: 'Never trust the artist. Trust the tale'. For Lawrence, the 'proper function of the critic' was 'to save the tale from the artist who created it' but here our contributors have varying degrees of room for both the teller and the tale – providing, of course, that the teller is able to be identified at all.[8]

This methodological breadth has the further benefit of leaving both the overt and latent content of the visual texts open for interpretation. This is all the more relevant with the more recent visual texts. With the earliest sources, the visual is all we often have due to the lack of literacy, the lack of verbally recorded history or, most recently, the lack of witnesses themselves. The more contemporary the text, the more likely, on the other hand, that both written material and personal testimony are available alongside the visual evidence. This would, at first sight, seem to diminish the relative importance of the visual. However, when examined alongside other forms of evidence, the visual can prove an unwittingly revealing counterpoint to the verbal and the individual. This is in no small part because in its popular manifestations it is often less profoundly self-conscious while at the same time being more collectively authored. It can be revealing, in other words, in other ways.[9]

All the chapters in this volume are united, however, in that they are methodologically interpretive. They accept that evidence is rarely self-evident and that there is a vital difference, therefore, between evidence and proof. Historians are rarely (if ever!) provided with the complete picture of the past, and so much of their work involves filling in the gaps between what is known and what might reasonably be presumed.[10] This may involve both interpolation and extrapolation, but it is clear that it must also involve interpretation. In her introduction to Sultan and Mandel's 'de-contextualized' collection of archive photographs first published in book form as *Evidence* in 1977, Sandra Phillips observes that there is 'no obvious clue to the book's meaning except in the pictures' and refers to the 'intrinsic meaning' of the pictures themselves.[11] The problem is, of course, whether meaning can ever be intrinsic. Such is the very nature of evidence, and the visual is no exception. This, it might be argued, constitutes much of the appeal of history as an intellectual pursuit: it has a fierce obligation to the evidence while remaining, at the same time, an inevitably creative process.

When the editors started putting together this volume, we made a clear decision not to make it a 'reader' compiled of important but already familiar extracts and essays. Rather, we decided specially to commission new chapters from authorities in visual media. They are drawn from the United Kingdom, the United States, continental Europe and Canada. Each author was commissioned to write on a particular medium, but within that brief the choice of examples, methodologies and underlying themes was up to them. Although contributors could choose their own illustrative case studies from any particular nation or culture, there should still be room for the reader to apply the underlying argument and approach to his or her own chosen research. Similarly, we were not prescriptive about any thematic issues that the authors should seek to extract. Although we expected gender, class and identity (for example) to emerge from a number of case studies, we were by no means insistent that they should – or that they should be limited to that. We were wary, in other words, of telling our contributors what they might expect to find. In the same way, readers should be free to discover both the familiar and the unexpected in their own research, both now and in the future. All we asked was that the contributors should be united in their belief in the importance of visual evidence, together with a conviction that the visual is not merely illustrative of the written word.

The first chapter following this Introduction addresses, as indicated earlier, perhaps the most venerable visual medium – painting. The task might seem insuperable, for the universe of painting is so vast and diverse. The author, David Morgan, rises to this challenge by laying down a few valuable principles that scholars may use, not merely to attempt to view paintings with understanding, but to exploit them as rich primary sources and to devise ways to utilize them to inspire and support their future endeavours. He then makes use of a few carefully selected examples to demonstrate his approach. This, in fact, establishes a pattern that will be found – albeit with significant freedom necessitated by both medium and essayist – in the subsequent chapters. Next comes a study of a feisty category of art work: cartoons and satirical prints. Although frequently provocative, they are often overlooked by scholars except as illustrations, an ironic oblivion from which Nicholas Hiley retrieves them. Showing both the technological and ideological progression of prints from two centuries, he demonstrates techniques of analysis and alerts readers to the promises and dangers of dealing with this genre.

Three chapters on photography follow. In the first of these, Douglas R. Nickel addresses some of the assumptions commonly made about photographs – such as the belief that they have unique authority as visual documents because they are produced by machines, rather than by a human process of creation – and shows that, to the contrary, the meaning of a photograph is substantially created by its viewers. The invention of photography as a technology, he asserts, accompanied another equally important innovation that we must come to terms with before we can understand the true authority and value of photographs: the invention photography *as an idea*. As Nickel develops it, the idea of photography turns out to be one that can engage the imagination and intellect on many levels. David Phillips then builds directly on this foundation by taking these precepts and applying them to the uses of photographs as documents. He explores the meaning of 'document' – like the idea of the photograph itself, not as simple as it seems – discusses the work of some noted photographers, and demonstrates the importance of the practice of documentary photography in the United States. Catherine Whalen then completes this trilogy of contributions by highlighting a category of images that could seem to be at the opposite pole from those intentionally created as documents or works of art. Her subject embraces photographs that may have been made casually or commercially – an enormous trove that ranges from snapshots to portraits and other genres. Often these items are not situated in archives or published works. They may even have been lost or discarded by their creators, later to be found by interested strangers. Her sensitive and penetrating case study of one such example offers a profound appreciation of the value of 'found photographs'.

We turn next to the moving visual image, encountering it first in its earliest environment – film – and then assessing the newer medium of television. Whereas paintings, prints and photographs are static objects that freeze a moment in time, removing it from the flow of history and life, motion pictures appear to possess a greater fidelity to reality, as they add the element of action. On the other hand, that trick of technology – for the pictures appear to move only thanks to the 'persistence of vision' – is only the first of many additional analytical challenges. Motion pictures are media as well as objects: generally intended to be seen by large numbers of viewers. Thus, as Luke McKernan describes the development of the particular genre of the newsreel, he shows that it was not merely a conduit of the news but a significant creator of it as well. Naturally, this requires a consideration of the meaning of the concept of 'the news' itself and the surprisingly varied ways it was communicated on film. McKernan surveys the history of the newsreels and encourages fresh thinking about the ways they offer documentation and function as discussions of both medium and message. The trajectory of Pierre Sorlin's treatment of documentary films moves, not surprisingly, along quite similar lines. Quickly demonstrating that the category 'documentary' includes a vast quantity not only of items but also of types, he raises useful questions and methodologies that must inform the scholarly approach exploiting this rich resource. Michael Paris anchors his discussion of the feature film in the voluminous interpretive literature this genre has inspired. His approach is both historiographical and multinational as he reviews the development of the narrative film, advocates greater respect for the value of cinematic dramas as primary sources, and suggests a number of considerations that scholars must weigh in their use.

The final three chapters focus on television. In these, the interpretive schema elucidated in the previous contributions are joined by the insights of communication studies. The 'news' again becomes a subject, inasmuch as it occupies a prominent niche in television programming. Cynthia Carter and Stuart Allan seek to probe behind the familiar patterns of television news and to highlight the issues arising from this particular type of journalism that are easily overlooked because of its highly polished presentation. Centring their attention on American and British news broadcasting, they demonstrate that these provide models widely imitated throughout the world. Then, turning to the apparently more fictitious cosmos of TV drama, Máire Messenger Davis investigates, by means of case studies built on intensive field research, the question of how realistic an understanding of contemporary culture can be discerned from popular programmes. Of course, no consideration of television can be complete without due attention to its commercial nature. Although the world of business interacts with all visual media, it is arguably most influential on television. In the final chapter, Jacqueline Botterill and Stephen Kline provide a wide-ranging discussion of advertising in order to show that, by bringing its 'representation of contemporary social life into the home', television gave advertising its most effective medium by far. In contrast to many scholars who have found it most convenient to study 'the frozen tableaux offered by magazine ads', they seek to provide an integrated approach to the analysis of advertisements in print, radio and television.

Although diverse in their backgrounds, disciplines, examples and approaches, the contributors to this volume unite here to present a coherent case for the study of the visual as evidence. It facilitates both the investigation of the past and the understanding of the present. The visual, it must be repeated, is not mere illustration. It is the evidence.

Notes

1 Walker Evans, 'The Little Screens: A Photographic Essay by Lee Friedlander with a Comment by Walker Evans', *Harper's Bazaar*, February 1963, 126–9.
2 James Agee, *Let Us Now Praise Famous Men* (Boston, MA: Houghton Mifflin, [1941] 1988), xi.
3 Interview with Harold Evans, *Booknotes*, C-SPAN, 7 February 1999.
4 David Morgan, *The Sacred Gaze* (Berkeley, CA: University of California Press, 2005), 25.
5 Morgan, *Sacred Gaze*, 26.
6 David Morgan, *Visual Piety* (Berkeley, CA: University of California Press, 1998), 208.
7 John Ruskin, *Modern Painters*, vol. 3 (London: George Allen, 1887), 278–9.
8 D.H. Lawrence, *Studies in Classic American Literature* (London: Martin Secker, 1933), 8–9. Lawrence is frequently misquoted as referring to the teller (and not the artist) and the tale.
9 For more on the latent content of visual and popular texts, see Richard Howells, *The Myth of the Titanic* (London: Macmillan, 1999) and *Visual Culture* (Cambridge: Polity, 2003), especially Chapter 4: 'Ideology'.
10 Robert F. Forth in his afterword to Sultan and Mandel's *Evidence* speaks of the difference between the 'circumstantial and the evident' and the need to fill in the gaps between what is clearly not there. See Larry Sultan and Mike Mandel, *Evidence* (New York: Distributed Art Publishers, [1997] 2003), n.p.
11 Sandra S. Phillips, introduction to Sultan and Mandel, *Evidence*.

1 Painting as visual evidence: production, circulation, reception

by David Morgan

Method and terminology

Art historians scrutinize patterns in the subject matter or content of paintings, and give special attention to its variation over time or from one artist to another. This method of study is called *iconography*. It works by looking for the relationship of continuity and change among images that share a theme or motif, reading the similarity and difference against one another as indications of artistic intention. Suppose, for example, we are looking at the image of a smiling human face. We recognize it as smiling because it resembles other smiling faces, so we interpret the face to mean happiness. Alter the direction of the line depicting the mouth and the expression of the face changes. The face then differs from the body of smiling faces and the difference tells us something else.

But that does not fully explain why we think we know that the image signifies happiness. We know happiness from the connection we quickly discern between this face and the connection of the countless smiling faces we have seen before to everything that commonly accompanies them – laughter, the pleasure of family, the fond company of friends and, of course, our own smiling faces in photographs and mirrors. It is evident that visual meaning relies on formula (all smiling faces), but also on personal experience to invest the formula with feeling. Not merely my experience, but that of others, who behold and share smiling faces with me. Humans are social animals and experience images as part of their sociality. They tend to organize what they see into systems of meaning, in this case, the rudiments of visual communication. A smiling face becomes a unit within a visual grammar, as it were, whose rules are more or less discrete and therefore learnable. Thus, visual meaning bears on the circumstances of a community of image users. The same smiling face will mean slightly or even dramatically different things to people separated by different ranges of experience, by different conventions and associations, by different cultural grammars of visual communication.

Yet the notion of a visual grammar informing the communicative function of images needs to be qualified by the fact that occasion makes a very big difference. The same smiling face may look irenic in a child's colouring book, but diabolic in the midst of a horror film. Context matters enormously in communication. The use and setting to which an image is put changes an iconographical interpretation of it. This suggests that the reception of an image generates meaning that may diverge from the intention of its maker. Images circulate widely and acquire meanings as they do so. Iconography as a method, therefore, must look beyond the simplicity of a sender-receiver model of interpretation, in which the image is a static, encoded message. Scholars do well to think about images as permanently unfinished sites of meaning-construction. Images are cultural objects with sometimes meandering biographies. Their meanings depend very much, if not entirely, on the audience or public that is looking at and making use of them.

Although this instability presents the scholar with a challenge, it also represents a keen opportunity for it means that images are part of the larger social life of communities and that they may, therefore, be used as evidence to tell us something about those among whom they circulate. An image is a social fact that may be applied as evidence to the task of historical or social analysis. Art historian Erwin Panofsky called this application *iconology*, that is, the study of the relationship between an image and the culture of ideas, values and other cultural forms that make up a world.[1]

In effect, iconography plots continuity and change in visual motifs while iconology discerns the significance of each by situating a particular image within the matrix of an entire age. Iconology looks for ideological connections between images and their social and cultural settings. And it does not assume that an image is engaged only in conveying conscious or intentional messages. One of the most fascinating powers of the image is its ability to work below the surface of consciousness. Images suggest affinities or sympathies to which viewers are attracted. Attentive readings of images by scholars can reveal ideological alignments that demonstrate how images may serve as cultural icons – windows onto a social world or historical moment that offer insightful evidentiary value for social and historical explanation.

Case study

As a case study of iconographical and iconological analysis, let us suppose that a cultural historian of the nineteenth century might be interested in forms of national consciousness in the wake of the American Civil War (1861–5) and that one might wish in this context to understand how Americans made use of their past as a way of understanding their present national identity. An especially destructive event in American history, the war controverted the nation's imagined community as a unified political and social entity. It was a catastrophic event that demanded deep considerations about what the nation now was and how it would go about affirming its mission. The search for a useable past is an enterprise common to communities in crisis. There is evidence of it everywhere. Just before and just after the war, Americans spent a good deal of time reading about and looking at representations of Puritans, and Pilgrims, the subset of Puritans who were distinguished by their opposition to the Established or state Church of England and suffered persecution for their illegal separation from the Church, and therefore sailed in 1620 to the New World to establish a colony that might allow them religious liberty.

Americans were fascinated, even obsessed with the Puritans. Sometimes the obsession could be quite dark, as in the case of Nathaniel Hawthorne's *The Scarlet Letter* (1850), which excoriated the hypocrisy of a Puritan community, but regarded the stigma that Hester Prynne bore as the very force that propelled her into a self-discovery that would never have occurred otherwise, as Hawthorne himself summarized: 'The tendency of her fate and fortunes had been to set her free. The scarlet letter was her passport into regions where other women dared not tread'.[2] But most evocations of Puritan history were more positive. During the same decade another widely received literary portrayal of Puritan times commanded popular admiration: Henry Wadsworth Longfellow's *The Courtship of Myles Standish* (1858), a love story that presented a strong portrait of a female hero, Priscilla. Over the course of the 1850s a number of historical works appeared. British author William Henry Bartlett (well known in America for several volumes of pious Holy Land pilgrimage literature) used what had became since the 1790s a commonly asserted patriarchal nomenclature in the title of his book, *The Pilgrim Fathers; or, The Founders of New England in the Reign of James the First* (1853).[3] Three years later a Boston publisher issued the long-lost journal of William Bradford, governor of Plymouth. Missing since the Revolutionary War, the journal offered annual records of the Pilgrim's city from 1620 to 1646. It was a bonanza for historians and sparked a new round of historical volumes and essays both popular and scholarly.

After the Civil War fascination with matters Puritan continued. In 1867, the American Tract Society issued a history aping Bartlett's title, *The Pilgrim Fathers of New England*, a volume directed towards pious libraries at home and Sunday school.[4] In the same year an Anglo-American artist named George Henry Boughton produced a small painting that would be widely reproduced and admired, *Pilgrims Going to Church* (Fig. 1.1). I'd like to focus on this painting as an exercise in using visual evidence.

Figure 1.1 George Henry Boughton, *Pilgrims Going to Church*

Boughton was born near Norwich in 1833, but emigrated to Albany, New York, as a very young boy, and grew up there. He exhibited his first painting in 1852 and showed two paintings at the National Academy of Design in 1853. He visited England, Scotland and Ireland briefly in 1856, then settled in London in 1862 after spending a year in Paris studying art. Boughton was elected a member of the National Academy of Design in 1871, associate of the Royal Academy in 1879, and member in 1896. He remained in London until his death in 1905.[5] As a painter of historical subjects conceived primarily as portraits, Boughton was well received on both sides of the Atlantic. He exhibited his work in major international exhibitions and received very favourable reviews.[6]

But none of his images captured the public's attention like *Pilgrims Going to Church*. Certainly all of his Puritan paintings participated in the ethos promoted by Hawthorne and Longfellow, and the title of his painting *Landing of the Pilgrim Fathers* even echoed Bartlett's widely read illustrated volume of popular history. But it only begs the question to frame Boughton's paintings in terms of the iconographer's habit of seeking out literary texts to reveal the programme and intention of the artist. If we take his work as no more than illustrations of Hawthorne, Longfellow or Bartlett, we still need to answer what Puritan subjects meant to them and their readers. In a long article on the painter that appeared in the year he was elected a member of the Royal Academy (1896), American educator, writer and Congregational minister, William Elliot Griffis, acknowledged Boughton's familiarity with Hawthorne and Longfellow, but saw all three as part of an American cultural movement:

> Roughly speaking, may we not say that the renaissance of the Pilgrim's story in art, prose literature and poetry dates from the discovery of Bradford's manuscript history, printed in 1856? After Hawthorne and Longfellow, Boughton deserves to rank as a true illuminator.[7]

Griffis would have noticed Boughton's Puritan themes for he shared the artist's fascination with Holland, Protestantism and early American history. Author of more than 40 books, Griffis was a robust contributor to what might be called the 'Puritan Renaissance' in the second half of the nineteenth century.[8]

So what did Boughton's often reproduced painting contribute to the era's fascination with early New England? To answer this iconological query, we must pursue two lines of enquiry – intention and reception. This means asking two questions: what the artist proposed to do in his work and what his audiences found there.

The evidence of style and content

We begin by asking what the artist intended. There are two ways to approach intention – on visual terms and on the witness of textual documentation. One always hopes for the convenient short cut of a letter, autobiography or reported conversation in which artists unambiguously reveal their intentions. But even if fortune should provide that convenience, it must be measured against the visual evidence of the artefact for intention operates at conscious as well as unconscious levels. The purely visual discernment of authorial intention, the matter of attending to the evidence of style, consists of asking what the manner of representation can tell us about an artist's purpose. Closely related to that question is another visual enquiry, which pertains to the content of the image: why do artists select and arrange their subjects as they do?

Beginning with the visual testimony of style, we observe that Boughton painted small or medium-sized pictures in oil in which the palette or range of colour was typically restricted to dark and light tones. 'He generally chooses', Griffis pointed out, 'to work subtly within the narrow limits of gray effects. These and his pearly and silvery hues lend themselves admirably to engraving'.[9] It was not an inadvisable motive: painters in Boughton's day might count on a steadier income from black-and-white reproductions of their work as engravings or lithographs, or as reproductions in magazines. Certainly such reproductive media circulated their work considerably farther than exhibitions alone might do. The tonal orientation of Boughton's work also corresponded to another consistent feature of it, indeed, served as a vehicle for it: the message-driven clarity that was amenable to the singular importance of title and theme for an American public much given to the literary taste in art, that is, its conformity to character, epic, melodrama and the reassuring glow of moral uplift. 'The legend under each picture', Griffis observed without a shadow of contempt, 'usually tells finely the story on canvas'.[10]

Drawing on sure draftsmanship and balanced compositions, Boughton's pictures consistently deliver their message. Nothing in the application of paint or choice of colour or distribution of forms urges the viewer to dwell on the painter's means of execution. *Pilgrims Going to Church* is not a painting about paint or the act of painting. The limpid application of pigment to the smooth surface, the silhouetting of the figures against the white background of a snowy field, and the lucid organization in bands of light and dark combine to create a bold pattern of tones that quickly deliver the gist of the scene. Everything bears on the silence of the moment. The style of the painting conditions the sort of experience one has beholding it and assures a legibility that never conflicts with the sense of watching an event from long ago. Boughton's way of painting facilitates access to that event, embracing the fiction of seeing what could not be seen and indulging the desire to do so. His was a style that rendered itself transparent in order to foreground access to the painting's content. The point was to conjure the illusion of a world long past and to place the viewer raptly before it. The kind of viewing that Boughton's picture invited might be said to have anticipated cinematic viewership.

From style we pass to the subject matter of the painting, that is, the image's contents – not *how* Boughton painted, but *what*. It is a procession of people in seventeenth-century dress, moving towards us from the left and then away towards the right. The company may be turning to their right, as if at the edge of a dense wood that begins in the three trees that separate them from us by marking the picture plane (the transparent pane separating the fictive space of the composition from the world of the viewer standing before the painting). If that is so, then viewers become an audience embedded in the woods, unseen by the Pilgrims, eye witnesses in the same patch of forest in which their enemies may hide, waiting to ambush the group. This gives viewers front-row seating and heightens their anticipation. The eye readily scans the length of the group, picking out their minutely depicted details – the fine, white collars, the boxy cut of leather footwear, the heavy fabric of their winter clothing, the dull gleam running the length of their rifles. Each of them clasps and conspicuously displays a thick book, the bibles they will use in worship.

The group is led by two bearded, armed men; behind them solemnly march the clergyman and his wife. His status is signalled not only by his position in the line, but also by the fact that he is one of only two men who wear gloves (the other is the armed man in front of him, dressed in a gold-brocaded garment, possibly indicating military rank, but more likely his social stature as a lawyer, as comparison with contemporary costume suggests).[11] The clergyman carries two volumes, evidence of his erudition and authority. Behind him are two girls, his daughters, perhaps, then more women and children accompanied by a helmeted guard. At the rear of the group are an older and a younger man, perhaps father and son. The young man has just halted, motioning his companion to hearken as they both glance nervously offstage, into the wood that is scantly signified by the sliver of a trunk that appears on the lower right corner of the canvas. The young man's bible is tucked in his belt, freeing his hand for action. At the same moment, a comrade follows many steps behind them. He is blissfully ignorant of the impending danger, his head lowered in the thoughtless rhythm of his gait.

The wealth of details assembled in Boughton's picture combines with the suspenseful composition of figures and the meticulous realism of his style of painting mentioned above to create a kind of historical window. The effect is important because it goes to the artist's intention as well as to the viewer's response. Viewers are made to feel as if they were eye-witnesses. As one reviewer said of the picture in the year it was first exhibited at the Royal Academy in London: 'It is truthful without ostentation, simple without guile'.[12] Boughton expended great effort on the details in order that their (at least apparent) accuracy would underscore the immediacy of the event about to unfold before viewers. He might have not bothered, as the illustrator of the same theme did not in the 1852 edition of *The Family Christian Almanac* (Fig. 1.2). A caption accompanying that illustration reads that 'The early settlers of New-England were often compelled to go to church armed, that they might be ready, if need be, to defend themselves in case of an attack by the Indians'.[13] But the coats, hats and gown are not seventeenth century at all, but eighteenth century.[14] And the round-topped windows were used by Anglicans in Virginia, not New England Puritans. Such details as these may not have mattered to the illustrator or the almanac's readership. Boughton, on the other hand, likely regarded the appearance of historical accuracy as necessary since it contributed to the dramatic success of the picture. But then accuracy is a relative matter. The log cabin dimly visible at the far

THE FAMILY CHRISTIAN ALMANAC.   23

The early settlers of New-England were often com-

Figure 1.2 Eli Whitney, Engraver, *Early Settlers of New England Going to Church*

left of the image was not the form of building used by the Puritans at this period. The guns appear to combine seventeenth-century firing mechanisms with later barrel design. These and other inconsistencies notwithstanding, however, his picture apparently struck contemporaries as a study in authentic detail.[15]

As a way of discerning the artist's intention, consideration of the painting's subject matter finds assistance in what little documentation we have. According to A.L. Baldry, who wrote the longest and best article on the artist's work one year before Boughton died, the scene was based on a passage from Bartlett's *Pilgrim Fathers*.[16] The claim appears to be corroborated by a handwritten note that was attached to the wooden panel that backed the painting. Since the note was written on paper bearing the monogram of the artist, it seems reasonable to assume that it originated with him. The note reads 'Early Puritans of New-England going to Church – armed to protect themselves from the Indians and wild beasts', which was the title in full listed in the 1867 review of the picture. Then the note continues with a selective quotation from the appendix of Bartlett's book:

> The few villages were almost isolated, being connected only by long miles of blind pathway through the woods; and helpless indeed was the position of that solitary settler who erected his rude hut in the midst of the acre or two of ground that he had cleared ... The cavalcade proceeding to church, the marriage procession – if marriage could be thought of in those frightful days – was often interrupted by the death shot from some invisible enemy.[17]

I say 'selective quotation' because it was not continuous, but elided several portions of Bartlett's text. Boughton condensed Bartlett's words in order to extract from them what interested him most: the isolation of the settlers and the menace of the public observance of any sacred rite subjected to ambush by enemies concealed in the dense woods of the Massachusetts wilderness. This is the bare iconography of the picture.

Was the painting therefore an illustration of Bartlett's text? Certainly the image conveys the sense of isolation, the precarious situation of the small band of Pilgrims. Boughton uses the right edge of the painting to suggest the ominous presence of 'some invisible enemy', all the while the two armed men who lead the group, at the other end of the painting, seem casually engaged in a conversation that leads them to relax the hold on their weaponry. The painter captures the sense of the momentary that Bartlett mentions, the imminence of a sudden interruption of gun shot. And he does it masterfully with the spacing of trees in the foreground: to the left appears a small interval occupied by the two guards forming the end of the group, straining to make out what they have heard. But the larger interval marked by two trees contains the bulk of the procession whose figures are glimpsed in another window of the same moment, most of them not having heard the sound – though a young girl in the centre glances to the right, suggesting the awareness of the menace may be spreading through the group. We witness in this painting the several facets of a single instant, all of which conspire to create a sense of *kairos*, the pregnant moment, which, in this case, may give birth to precipitous action and the irreversible entropy of violence.

From this examination of style, subject and documentation, we may offer reasonable inferences regarding the artist's purpose. It helps us imagine the intended effect of the picture: like its protagonists, viewers do not see what threatens the pious troupe. The enemy is as unseen and distant as the Pilgrims are unconcealed and close to us. The threat was compounded by the fact that, as Bartlett points out, the Indians had 'gradually come to obtain possession of fire-arms' by the time of King Philip's War (1675–6).[18] This informs the theatrical staging of Boughton's painting.

The evidence of reception: from iconography to iconology

Discerning the intended effect of the painting advances the investigation to a second domain of enquiry – why the selection and staging of this moment appealed to Boughton's audience. Answering this question will address the task we have set ourselves: how the painting might contribute as evidence to understanding what Pilgrims meant in Victorian America. In 1879, a writer drew attention to the 'suspense of apprehension' struck by Boughton's picture, whose silence 'may instantly undergo a transformation'.[19] In 1883 a critic praised Boughton for showing 'sufficient dramatic power for a strong theme, as witness his "Pilgrims going to Church"... and the gloom of Hester Prynne on a mission of mercy to a house stricken with the plague'.[20] The picture also came to signify something epochal for many. Citing the painting, William Elliot Griffis proclaimed in 1896 that for nearly three decades Boughton had become 'the interpreter and we may say illuminator of New England life in the seventeenth century'.[21] Baldry characterized the painting as 'an illustration to the chronicles of a country which owes the commencement of its prosperity and progress to the indomitable spirit of the first settlers in districts where almost everything conspired to hamper the spread of civilization'. It was an image baring the righteous roots of America. And it was an ideological icon, a window on the historical reality that grounded the present. The small company is America in microcosm, banded together against unseen threat. Baldry praised the 'historical value' of the artist's Puritan pictures: 'they show, with a dramatic directness ... what were the conditions under which the foundations of the American Republic were laid'.[22]

But another way to understand what Boughton's painting meant to his contemporaries asks what visual evidence can tell us. And so we pass from iconographical investigation to iconological. But the results of our enquiry thus far are not cast aside; we may place them in escrow for the moment. We already noted the remark by Griffis that Boughton's tonal compositions accommodated graphic reproduction. This fact helps account for the visual reception of *Pilgrims Going to Church*: it was reproduced on many occasions over the decades following its creation.[23] (The painting was so popular that the artist even produced a smaller oil version of it in 1872, which went to a British patron in Boughton's day, but is now in the collection of the Toledo Museum of Art.[24]) The frequent reproduction of the painting provides a visual record of reception since we may glean from each instance some notion of why the image was being reproduced, and therefore what it meant on each occasion as an evocation of seventeenth-century America.

In January 1893, the *New England Magazine* ran a reproduction of Boughton's painting following an editorial review of a new edition of John Greenleaf Whittier's Complete Works. No comment or caption provides a hint about why the image was there. No other item in the issue used it as an illustration. But the long essay on the three volumes of Whittier's work dwelled on his treatment of English Puritans and Quakers and American Puritan history, particularly their resistance to an unjust government.[25] Why was Boughton's picture there? It had become a kind of icon of what Griffis wrote was the painter's reputation for portraying 'the more lovely side of the Pilgrims and Puritans'. (It was Hawthorne, he said, who 'leads us into the dark shadow and the terrible gloom'.[26]) Loveliest about the Pilgrims was their yearning for liberty, something Whittier scolded latter-day American Protestants for overlooking: 'It has been the fashion of a class of shallow Church and State defenders', he is quoted in *The New England Magazine*, 'to ridicule the great men of the Commonwealth, the sturdy republicans of England.' Perhaps *Pilgrims Going to Church* was meant to counter that tendency. It certainly became that in the history of the painting's reception. In 1920 a Harvard government professor gathered five of Boughton's paintings, including *Pilgrims Going to Church*, to illustrate his article on 'The Pilgrims', and to champion Boughton's treatment of a theme near to his heart and his audience's:

> It remained for George Henry Boughton to find something in the life of the Pilgrims besides misery and gloom, and to picture them in ways that are endearing and inspiring to those of us today who look back to our Puritan forbears with proud appreciation of their splendid courage, their simple, upright character, and their steadfast devotion to the cause of Liberty. Boughton has caught and reflected sympathetically the spirit of that struggling little community of Plymouth, and has shown us how they lived, loved, worked, and worshiped ... Boughton's art is identified in our minds with Puritan New England.[27]

But this view did not last. A few years later the painting was reproduced in the first volume of a popular series called The Pageant of America, where it was used to stress the densely intertwined religious and political life of the Puritans. The caption read in part:

> The little Massachusetts towns that sprang up about Boston were each centered in the church. No one could be a freeman and participate in the political life of the colony who was not a church member, and the membership was carefully restricted and supervised. Rigid laws compelled all to support and attend the institutions of worship.[28]

Note that all mention of the Indian menace has fallen away, replaced by the compelling civil authorities derived from what the caption called 'Calvin's ecclesiastical government'. The scene no longer portrayed the bravery of pious Pilgrims, but the civic necessity and legal coercion of church attendance. The other, Hawthornian view of Puritanism was rising. If Americans admired the quest for liberty that brought the Pilgrims to the shores of Massachusetts in the first place, many were not prepared to celebrate the Puritan intolerances that emerged in the treatment of Quakers and Baptists, or in the heterodoxy of Anne Hutchinson or the Salem witchcraft crisis. The volume opened with a 68-page prologue on 'The American India', followed by chapters on the Vikings, Portuguese explorers, Christopher Columbus, the Western Route to Asia, New Spain and Jamestown. The Puritans do not appear until page 193. By 1925, the popular national narrative had caught up with the professional historian's parsing of continental (rather than merely national) history, which recognized the diversity and considerable age of origins, and therefore found less use for the scions of New England.[29] Though it continued to appear from time to time, the cultural biography of Boughton's picture may have passed its zenith.

These and many other instances of the reproduction of *Pilgrims Going to Church* trace a long practice of regarding this as well as other pictures by the artist as a reliable window on early New England life. Americans have remained enthralled with Hawthorne's reading of the Puritan past, as shown by Arthur Miller's *The Crucible* (1952), portraying the feverish nature of collective hysteria and political terror, or even M. Night Shyamalan's film inspired by the play, *The Village* (2004). But Victorian Americans also wanted another register in which to imagine the founding moments of the colonial past, one in which to regard Puritanism as a positive origin. And this is precisely what Boughton's pictures gave them. The Puritans were not all bad. But why was it compelling for Americans to imagine this?

There is another, distinctly visual way of answering this question. We can understand the importance of *Pilgrims Going to Church* as part of a larger pattern of feeling, memory and association by comparing it with images of its day which share a significant aspect of its theme, yet also bear a meaningful difference. It is in the difference that we may discover further clues to the received if not intended meaning of Boughton's picture. As will become clear, Boughton's picture was part of a larger visual discourse, about which he may or may not have been cognizant. There was a contemporary visual setting in which his work was seen, a thematic ecology or iconology in which his work participated.

Had Boughton wanted to portray no more than the popular theme of church going, he might have done so in the picturesque form of the style popularized by George Henry Durrie (Fig. 1.3). Durrie was interested in the American past too, but not the historical past. His work was loved and collected for its ability to call up a nostalgic vision of a cosy rural America irradiating a pleasant reverie, an American Biedermeier simplicity, rustic, but not without delicacy and the *civilitas* of well-crafted clapboard architecture, the industry of yeomen, and the comfort of polished sleighs and well-managed horse teams. A welcoming, picturesque landscape unrolls to nest a gleaming bit of civilization. The landscape is dominated by the massive church, which dwarfs the few buildings about it. The steeple rises higher into the luminous sky than the trees and densely wooded valley walls, proudly proclaiming a covenant between heaven and the rustic nation that hustles dutifully to worship on a winter Sabbath.

The same theme was treated in a print by Philadelphia lithographer Augustus Kollner about 1850, which was issued by the American Sunday School Union (Fig. 1.4). Entitled *The Happy Family*, the image shows a group departing their home one Sabbath morning to make their way through the light woods to the church, whose spire looms in the distance. The composition bears comparison with Boughton's because it arrays the family in a manner that anticipated his construction of the Pilgrim company: bounded by protective males clasping the Holy Book, the woman and children occupy the safe interior of the small procession. One surmises that Boughton's image may have appealed to contemporaries not only because of its paternalistic structuring of gender, but also because his painting is a visual argument that the same structure is rooted in the origins of the nation. By protecting their children and women folk, Victorian men were acting in

Figure 1.3 George Durrie, *Going to Church*

accord with the oldest and most pious tradition of their forebears, the legacy of the 'Pilgrim Fathers'.

The Sunday School Union print sheds further light on the deceptive emotional temperature of Boughton's protagonists. A caption informs us that the 'happy family ... with neat attire and cheerful hearts ... go up to the courts of the Lord'. The neat attire is clear enough, but we must take the caption on faith regarding 'cheerful hearts' for nothing in the image betrays them. No smile or light touch, no spring in the gait. Their solemn demeanour is matched by the clumping of heavily cloaked bodies. But this restrained expression may very well temper our reading of Boughton's Pilgrims, who seem no less sombre. Presumably the gravity suits the occasion. The Sabbath is no light-hearted matter. Yet Kollner's lithograph may correct our response to Boughton's portrayal of the Pilgrims whose restraint we may be apt to read as humourless and gloomy. Victorian Americans may not have seen it that way. What art historian Michael Baxandall called 'the period eye' may have read in this instance the rich subtleties of grey, brown, buff and green clothing, studded with the syncopated rhythm of immaculately white collars, as a material cheerfulness of heart en route to the high point of the Pilgrim's weekly sojourn on earth – Sabbath worship.[30]

Another contemporary picture is closer to Boughton's fascination with historical drama and is freighted with the national sense of purpose that corresponds to the ongoing reception of *Pilgrims Going to Church*. In 1852, George Caleb Bingham completed his major canvas, *Daniel Boone Escorting Settlers through the Cumberland Gap* (Fig. 1.5),

THE HAPPY FAMILY.

The happy family are on their way to the place of public worship It is Sunday morning, and with neat attire and cheerful heart they go up to the courts of the Lord

Figure 1.4 Augustus Kollner, *The Happy Family*

confident, as he wrote to a friend, that 'the subject is a popular one in the West, and one which has never been painted'.[31] He was so confident that this image would appeal in the present moment of westward expansion in the wake of the war with Mexico (1848), which dramatically extended national territory, that he wondered whether he should sell the painting outright or 'have it engraved with the expectation of remunerating myself from the sale of the engraving'.[32] Bingham cast Boone in a role that the historical figure often performed: leading easterners into the frontier where they might settle as farmers. Although the famed Indian fighter pictured at the head of the procession does not escort the company to church, Boone does lead the pilgrims on a sacred mission. The band of American Israelites follows a homegrown Moses, who was legendary during his own lifetime. Bingham chose a dramatic composition that places the viewer immediately before the exodus, whose elect emerge from gloom into a sharply defining light that also etches the blasted stumps and branches to either side of the company, as if a terrifying, providential force had cleared the way for their passage. That force was in fact – in the allegory of the expanding nation – President James Polk, who seized land from Mexico through armed exploit. Bingham's picture sanctifies violence by situating it within a larger national narrative.

Boughton's painting takes on a different tenor when placed within the chorus hymned by pictures like Figs. 1.2 and 1.5. Although many Pilgrims and Puritans were

Figure 1.5 George Caleb Bingham, *Daniel Boone Escorting Settlers through the Cumberland Gap*

killed by Indians in King Philip's War, rather few Americans in the late nineteenth century were besieged by Indian attacks. True, Plains tribes during the 1870s had rebelled against the government, forsaking the reservations to which they had been assigned, most spectacularly under the leadership of Chief Joseph of the Nez-Perces during the summer of 1876, when General Custer met his infamous demise and just as the Centennial Exposition, which included *Pilgrims Going to Church*, was opening in Philadelphia. But the larger menace facing Americans generally speaking was the aftermath of the Civil War and the challenges of economic and political reconstruction. Boughton never painted anything that had to do with the Civil War. He had left the nation before the war had broken out and never returned. But we are not talking about what the artist thought or intended. The question at hand, a matter of iconological interpretation, is why Americans for so long found his picture of *Pilgrims Going to Church* so appealing. Griffis said something suggestive at the outset of his 1896 essay on the artist:

> Since that ancient and historic epoch, 'the war', George Henry Boughton's works have made for their author a name and a place in American homes as that of the painter of early New England life and, especially, of the more lovely side of the Pilgrims and Puritans.[33]

It all began *after* the war, as if the icons of national origin helped to forget the carnage and the tear in the heart of the nation. This comports with Lincoln's elevation of

Thanksgiving to a national holiday in the midst of the war, in 1863. The holiday was intended to envision a happier day, the origin of a national unity now violently ripped asunder.

To the war one might easily add another new feature of American life since mid-century: the rising numbers of non-English-speaking immigrants arriving every day, so many that the originally Puritan capital of Boston became a centre of Roman Catholicism. In light of these two singular changes in the nation, the vision of Pilgrims banding together against the threat of imminent attack may have evoked a moment of unity, a communal mythos that would prevail against the threats of annihilation that it faced. For a time, Boughton's Pilgrims became a cultural icon to many Americans because it answered to the broad civic piety that was devoted to 'the Pilgrim Fathers'. Boughton called the Fathers forth in a totemic image that bridged their time and the painter's own. That must have been a comforting message as Americans struggled to determine what the Civil War meant and what the changing ethnicities of new Americans portended. The picture counselled facing a frenzied future by anchoring the nation in what appeared to be the deep waters of a stable past.

Acknowledgement

My thanks to Gretchen Buggeln for her perceptive comments on an earlier draft of this chapter.

Notes

1 Erwin Panofsky, 'Iconography and Iconology: An Introduction to the Study of Renaissance Art', revised and reprinted in *Meaning in the Visual Arts* (Garden City, NY: Doubleday Anchor, [1939] 1955), 26–54. For an instructive discussion of iconology as it relates to the study of visual media, see Richard Howells, *Visual Culture* (Cambridge: Polity, 2003), 11–31.

2 Nathaniel Hawthorne, *The Scarlet Letter* (New York: Barnes & Noble Classics, 2003), 165.

3 W.H. Bartlett, *The Pilgrim Fathers; or, The Founders of New England in the Reign of James the First* (London: Arthur Hall, Virtue, 1853). For a study of the history of the term, see Albert Matthews, *The Term Pilgrim Fathers* (Cambridge, MA: John Wilson and Son, University Press, 1915; reprinted from *The Publications of the Colonial Society of Massachusetts*, vol. 17, 293–391), who argued (p. 352) that the term first appeared in 1799.

4 W. Carlos Martyn, *The Pilgrim Fathers of New England: A History* (New York: American Tract Society, 1867).

5 A good source of information about the painter is Alfred Lys Baldry, 'George H. Boughton, R.A., His Life and Work', *The Art Annual* (Christmas number of *The Art-Journal*, 1904): 1–32. A perceptive and well-researched study of the reception of Boughton's painting is Amanda J. Glesmann, 'Sentimental Journey: Envisioning the American Past in George Henry Boughton's *Pilgrims Going to Church*' (MA thesis, University of Delaware, 2002).

6 Several of his Puritan New England themes are reproduced as steel engravings in William Elliot Griffis, 'George H. Boughton, The Painter of New England Puritanism', *The New*

England Magazine 15, no. 4 (December 1896): 481–501; several more, along with exhibition lists, appear in Baldry, 'George H. Boughton'. For a discussion of several paintings of Puritan subjects by Boughton, in particular the often-reproduced *Pilgrims Going to Church*, see Roger B. Stein, 'Gilded Age Pilgrims', in *Picturing Old New England: Image and Memory*, ed. William H. Truettner and Roger B. Stein, exhibition catalogue, National Museum of American Art, Smithsonian Institution (New Haven, CT: Yale University Press, 1999), 45–7.

7   Griffis, 'George H. Boughton', 495.

8   Griffis published several volumes on Puritan history on both sides of the Atlantic, including *The Romance of American Colonization: How the Foundation Stones of our History were Laid* (Boston, MA: W.A. Wilde, 1897); *The Pilgrims in their Three Homes, England, Holland, America* (Boston, MA: Mifflin, 1898; rev. ed., 1914); *Young People's History of the Pilgrims* (Boston, MA: Houghton Mifflin, 1920).

9   Griffis, 'George H. Boughton', 491.

10   Ibid.

11   The gold brocade trim terminating in buttons resembles a lawyer's coat during the brief reign of James II (1685–8), as pictured in Elisabeth McClellan, *History of American Costume 1607–1870* (New York: Tudor, 1937), 120.

12   Anonymous review, 'The Royal Academy', *The Art Journal* 6 (1 June 1867): 140.

13   *The Illustrated Family Christian Almanac for 1852* (New York: American Tract Society, 1851), 23.

14   See McClellan, *History of American Costume*, 156, 166, 221. The three-cornered cocked hat was exclusively an eighteenth-century fashion.

15   For further discussion of inaccuracies, see Glesmann, 'Sentimental Journey', 6–7.

16   Baldry, 'George Henry Boughton', 8.

17   Quoted in Richard J. Koke, ed., *American Landscape and Genre Paintings* (New York: New York Historical Society, 1982), 1: 77. Compare Bartlett, *Pilgrim Fathers*, 236–7. Much of the appendix consisted of Bartlett quoting his own earlier book, *The History of America* (Boston, MA: Samuel Walker, 1843).

18   Barlett, *Pilgrim Fathers*, 236.

19   Edward Strahan, 'The Collection of Mr. Robert L. Stuart', in Strahan, *Art Treasures in America* (Philadelphia, PA: George Barrie, 1879), 118.

20   James D. McCabe, *The Illustrated History of the Centennial Exposition* (Philadelphia, PA: National Publishing Company, 1876), 524; Joseph Hatton, 'Some Glimpses of Artistic London', *Harper's New Monthly Magazine* 67, no. 402 (1883): 842.

21   Griffis, 'George H. Boughton', 495.

22   Baldry, 'George Henry Boughton', 11. For further consideration of the picture's critical reception, see Glesmann, 'Sentimental Journey'.

23   For a listing of several of these instances of reproduction, see Stein, 'Gilded Age Pilgrims', 46 and 73, n.13.

24   The title differs: *Early Puritans of New England Going to Worship*, Toledo Museum of Art, accession number 57.30. It is reproduced in Susan E. Strickler, *American Paintings*, ed. William Hutton (Toledo, OH: Toledo Museum of Art; University Park, PA: distributed by the Pennsylvania State University Press, 1979), 140.

25   'Editor's Table', *The New England Magazine* 7, no. 5 (January 1893): 674–80.

26   Griffis, 'George H. Boughton', 481 and 498.

27 Albert Bushwell Hart, 'The Story of the Pilgrims', *The Mentor*, 8 (November 1920): 20–1.

28 Clark Wissler, Constance Lindsay Skinner and William Wood, *Adventurers in the Wilderness* (New Haven, CT: Yale University Press, 1925), 212.

29 American historians had long stressed the complexity and diversity of continental history. See, for example, George Bancroft, *History of the United States of America: From the Discovery of the Continent*, 4 vols (Boston, MA: Little, Brown, 1846–75).

30 Michael Baxandall, *Painting and Experience in Fifteenth Century Italy: A Primer in the Social History of Pictorial Style* (Oxford: Oxford University Press, 1972), 29–32.

31 Elizabeth Johns, 'The "Missouri Artist" as Artist', in *George Caleb Bingham*, exhibition catalogue, Saint Louis Museum of Art, ed. Michael Edward Shapiro (New York: Harry N. Abrams, 1990), 136.

32 Ibid.

33 Griffis, 'George H. Boughton', 481.

2 Showing politics to the people: cartoons, comics and satirical prints

by Nicholas Hiley

The use of satirical prints and cartoons in academic books, articles and lectures is strangely limited, as it is almost entirely confined to their visual content. Prints and cartoons are taken as evidence only of what they show, and are generally used solely in relation to the individuals and events they depict. There is seldom any reference to the fact that they were created by a specific artist, using a particular reproductive process, for a known publisher, and an identifiable readership. If details of the artist or original publication are included with the image, it is often because the copyright holder has insisted on it.

This is a bleak situation, but it is made worse by the fact that the majority of these prints and cartoons are used solely as illustration. They are included only because they depict individuals or events mentioned in the accompanying text, and are seldom an important part of the argument, which continues to rely for its force on written, and not pictorial, evidence. Moreover, in many cases the choice of these illustrations has been determined by their reproduction and copyright costs, and the final selection left to the publisher – something unthinkable for any other part of an academic text.

The neglect of this type of visual evidence goes further. Because academics predominantly use satirical prints and cartoons as illustrations, they select those most easily understood out of context. The original artist worked deep in context, following changing political events from day to day, attending to specific editorial demands, and acutely aware of shifts in public opinion. But most academics choose only those images that are instantly recognizable to a modern audience, with no detailed contextual explanation. A surprising number of academics in fact photocopy prints and cartoons without noting either the artist or the source, and leave their publisher to chase these details for them.

This attitude affects all forms of visual evidence available to academics. It results in the downgrading of these prints and cartoons, through the neglect of the contexts in which they were created, distributed and received. It is perhaps understandable, for it relieves academics of the necessity for analysing the complex industries which created these images; the processes by which they were made and printed; the relationship between the cartoonist and the publication for which they worked; the circulation and readership of that publication; and the impact of those cartoons upon readers. Yet it is in those rich complexities that the principal value of this evidence lies.

The best way to demonstrate this rich set of contexts is to look at a series of satirical prints and cartoons spanning 200 years of graphic satire, to examine the mass of information they carry, all of it evidence of the dynamic relationship between public figures, the media and their audiences, and the technologies that bind them together.

Figure 2.1 James Gillray, 'Dido in Despair'

The first image is of a print by the 44-year-old James Gillray, published in London on 6 February 1801, and entitled 'Dido in Despair' (Fig. 2.1).[1] Its subject was the notorious affair between Horatio Nelson, the naval hero, and Lady Emma Hamilton. The caption refers to the classical legend, in which Queen Dido killed herself with grief when Aeneas sailed away from Carthage. Gillray shows Emma leaping out of bed in her nightgown, leaving her husband sleeping. Through the window is the fleet carrying Nelson to his next command, and the verse below reads:

> Ah, where, & ah where, is my gallant Sailor gone?
> He's gone to Fight the Frenchmen, for George upon the Throne.
> He's gone to fight ye Frenchmen, t'loose t'other Arm & Eye,
> And left me with the old Antiques, to lay me down & Cry.

The meaning of the print lies in the detail, and readers were expected to tease it out. The 's' on 'Antiques' is deliberately small and faint, emphasizing that the actual source of Emma's tears is her husband – 'the old antique'. Sir William Hamilton was a wealthy collector of classical antiquities who had just celebrated his 70th birthday, and Emma, only 35, was regarded as part of his collection – Horace Walpole joking that Hamilton 'has actually married his gallery of statues'.[2] Hamilton liked to show off his wife, and even acted as master of ceremonies while she posed in classical dress in imitation of the figures on his Greek vases. These were known as her 'Attitudes', and Gillray has drawn an open book on the window seat entitled 'Studies of Academic Attitudes taken from the Life'. However, the book illustration in fact shows Emma Hamilton naked and flat on her back, which many people did regard as her most characteristic attitude.[3]

As Gillray and his readers knew, Emma's affair with Nelson had begun in Naples, where her husband was envoy, and in November 1800 they had all returned to England together, in a relationship Emma called 'tria juncta in uno'. This seriously damaged Nelson's reputation, and gave Gillray the opportunity of creating this scurrilous satire. He clearly took enormous pleasure in every comic element, drawing the celebrated Emma as a self-dramatizing figure, who was in fact losing her beauty and putting on weight. This was all true, although in emphasizing her bulk he may have been unaware that Emma had just given birth to Nelson's child, in great secrecy.[4]

Gillray plainly enjoyed making fun of the elderly Hamilton, publishing two further cartoons of him later that same year. They were all etchings, prepared by Gillray himself from a preliminary sketch. In this process a copper printing plate was coated with a thin layer of protective varnish or wax, onto which the sketch was transferred. Gillray would then scratch the lines of the cartoon through this layer with an etching needle, working not only on the picture but also the lettering, which had to be drawn back to front. The plate was then immersed in acid, which ate away the metal exposed by the needle, to create the etched grooves in which the ink would sit for printing.[5] It was a slow process, which further encouraged the elaboration of detail.

The printing was done one sheet at a time on a wooden hand press, and it seems likely that the publisher – Hannah Humphrey – operated at least two such presses, possibly in the basement of her London shop. Two presses could turn out about 40 sheets an hour, and the several hundred copies of the first edition of a print could thus be produced in a few days. These would be offered for sale at a shilling each, or half-a-crown if coloured, the colouring probably added by a team of women following a master copy coloured by Gillray.[6] Demand for a new design could be high, and Gillray's 'Political Dreamings!', published a few months after this print, was so popular that the entire run sold out within days.[7]

It is difficult to estimate the size of the readership for these prints, but equally hard to imagine a process whereby images that were etched, printed and coloured by hand in short runs, and sold for relatively high prices, could have reached a significant proportion of the population, even in London.[8] We know that there was some sharing of prints, but their display in print-shop windows, and the hiring of portfolios, cannot have much increased their circulation. They were simply too expensive to circulate freely in coffee-shops and taverns, or to be displayed on their walls.[9] These were metropolitan satires, shared by a wealthy social group, and requiring the reader to have a detailed knowledge of the news and gossip circulating within it.

Gillray's print has to be read with this in mind, for it explains how every object in Emma's room can bear a comic message, from the discarded garter on the floor inscribed 'The Hero of the Nile', to the cosmetic pots on the dressing table (one labelled 'Rouge a la Naples') and the collection of Hamilton's antiquities on the floor. These include a bust of Messalina, the notoriously unfaithful wife of Emperor Claudius, representations of Venus and a satyr, and a very phallic statue which was once labelled 'Priapus', but is now broken and useless. This object was both a sly personal dig at Hamilton, and a joking reference to his supposed discovery, 20 years earlier, of a surviving Cult of Priapus in Italy. Every detail was etched by the cartoonist himself, and was there to be read if you had heard the gossip.

This type of small-circulation, highly informed print was the dominant type of visual satire in Britain during the first quarter of the nineteenth century. However, over the next 20 years the narrow market for printed satires began to expand with the extension of the franchise, and their distribution was transformed by the creation of illustrated comic magazines such as *Figaro in London* in 1831, and *Punch* in 1841. The second image is characteristic of this new development. It was published on 8 June 1867 in the fifth issue of a magazine called *The Tomahawk: A Saturday Journal of Satire*, and was by Matt Morgan, its 31-year-old principal cartoonist (Fig. 2.2).

WHERE IS BRITANNIA ?

Figure 2.2 Matt Morgan, 'Where is Britannia?'

This black-and-white cartoon addressed the matter of Queen Victoria's continued isolation since the death of her husband Albert, more than five years earlier. Captioned 'Where is Britannia?', it showed an empty throne, abandoned regalia and a sleeping lion. An accompanying open letter suggested that it was selfish for the Queen to be in mourning for so long, and that she had 'no right' to hide in Scotland, avoiding visiting heads of state, while entertaining 'Scotch ghillies and their families'.[10] This passage was a reference to the Queen's close friendship with her Scottish servant John Brown, which it seems had already led to the nickname 'Mrs John Brown'.[11]

The whole process of production and distribution of graphic satire had changed since Gillray's time. *The Tomahawk* cartoon was not an etching but a wood engraving, produced by Morgan in conjunction with the well-known wood engraver Thomas Bolton.[12] Morgan would have drawn his cartoon directly on the whitened face of a wooden printing-block, which consisted of several pieces of boxwood bolted together at the back. Once Morgan had finished his drawing, the bolts would have been unfastened, and Bolton would have distributed the separate sections to different wood engravers. They painstakingly cut away all the white areas of Morgan's drawing, leaving the black lines raised to carry the ink. The reassembled block was then combined with type to produce a page of text.[13]

Wood engraving was a painstaking process taking many hours, but it vastly increased the number of prints that could be made, and supported numerous satirical magazines. The market indeed became carefully subdivided. On its launch in May 1867 *The Tomahawk* was priced at 2d., placing it between *Fun*, established in 1861 and selling for 1d., and *Punch*, the market leader, launched in 1841 with a cover price of 3d. There was hardly room for another publication, but May 1867 had nevertheless also seen the launch of *Judy*, deliberately undercutting *The Tomahawk* at 1½d. Weekly sales are hard to estimate, but *Fun*, with its low cover price, probably needed a circulation of 10,000–20,000 to survive. *The Tomahawk*, with its higher cover price, could probably have survived on 10,000 copies a week, and may have been selling more, although it was unlikely to have reached *Punch*'s weekly circulation of around 40,000 copies.[14]

The different comic magazines addressed different readerships, largely defined by the cover price. Most readers would have come from the urban middle classes – with *Fun* probably addressing lower-middle-class readers, and *Punch* the upper-middle and upper classes. The sharing of old copies, and communal reading in pubs and coffee-houses, might have brought in less prosperous readers, but the range of subjects and style of address was still firmly defined by middle-class sensibilities. Readers of *The Tomahawk* were expected to understand references in French, German and Latin, and even Morgan's 'Where is Britannia?' assumed some knowledge of current political gossip.

Politics also played a part in subdividing the market. *Fun*, for example, supported the Liberal Party, as did *Punch* to a lesser extent. *Judy* supported the Conservatives, while *The Tomahawk* regarded itself as above party politics, treating it as an undignified struggle for power and office. This was evident in the subject of its cartoons, for *The Tomahawk* printed fewer than half the number of political cartoons carried by *Fun* or *Punch*, but twice as many cartoons on social issues.[15] The targets of *The Tomahawk* cartoons were decided at staff meetings, the week before publication, but Morgan had more freedom and greater space than cartoonists on rival publications, including double-page colour cartoons.[16]

The Tomahawk was keen to distinguish itself in a crowded market, and on 10 August 1867 Morgan returned to the subject of the monarchy with a large fold-out cartoon, in which the British lion had awakened, but Brown stood in front of the empty throne, preventing it getting to the crown. Captioned 'A Brown Study!', this caused a sensation and helped push the magazine's circulation to 50,000 copies.[17] This meant a total readership closer to 500,000, for journals like *The Tomahawk* were printed on durable rag-based paper, with each copy being shared and passed around for weeks or even months. 'A Brown Study!' became widely known, and when it was reported in October 1867 that a penny serial was being sold in London, entitled *John Brown, or the Fortunes of a Gillie*, the cover was said to show him 'in the famous attitude made known to all England by the *Tomahawk* cartoon'.[18]

'A Brown Study!' was notorious, and, after Morgan left to work in the United States in 1870, it was even said to have 'made London too hot to hold him'.[19] This was an exaggeration, but it emphasizes how radically different Morgan's cartoon is from Gillray's, not only in the way it was produced and distributed, but also in the number of people who saw and remembered it. By the final quarter of the nineteenth century there had been further changes, however, and these are embodied in the third image, which appeared on 12 April 1890 in the comic magazine *Ally Sloper's Half-Holiday*. Drawn by the 27-year-old William Thomas, it shows the eponymous Ally Sloper chatting pleasantly with the 74-year-old Otto von Bismarck, who had finally resigned as German Chancellor the previous month, after disagreements with Kaiser Wilhelm II (Fig. 2.3).

Figure 2.3 William Thomas, Ally Sloper

By the time Thomas' cartoon appeared, the cost of magazine production had fallen dramatically, as rag-based printing paper gave way to much cheaper alternatives based on wood-pulp. The cost of illustration had also dropped sharply, as cheaply produced 'process' blocks replaced expensive wood engravings. These metal printing plates were produced photographically from an original drawing, the image being transferred to the sensitized surface of the metal plate where, after careful treatment, the blank areas were etched away to leave the lines as a raised printing surface. The cartoonist no longer had any involvement in the blockmaking, and the process was quick and cheap. When combined with cheap paper it enabled the production of the first inexpensive, fully pictorial, mass-circulation magazines.

Graphic satire rapidly extended into cheap publications like *Ally Sloper's Half-Holiday*. The magazine itself claimed a circulation of 340,000 copies by the time this cartoon appeared, and was said to have remained around this mark, occasionally rising as high as 500,000 copies. This was a formidable figure, which indicated a readership of millions. Internal evidence, including the lists of competition winners, shows great popularity among lower-middle-class clerks, tradespeople and shopworkers, although it seems that the magazine also had large numbers of working-class readers. Once again, communal reading would have taken *Ally Sloper's Half-Holiday* to those who had been priced out of buying it.[20]

Ally Sloper's Half-Holiday was created by process engraving, but in fact owed its origins to a prominent firm of wood engravers, Dalziel Brothers. The magazine's proprietor, Gilbert Dalziel, had served his apprenticeship on the family-owned *Fun* before moving in 1872 to *Judy*, which his father bought that year. *Judy*'s editor was Charles Ross, a play-wright and former theatre manager, who had already created the disreputable figure of Ally Sloper.[21] As developed by Ross's cartoonist wife, Marie Duval, Sloper's name fitted his seedy nature, and his appeal was such that in 1873 the first of a series of spin-off books appeared, entitled *Ally Sloper: A Moral Lesson*.[22] Eventually, in May 1884, the Dalziels decided to capitalize on Sloper's success by launching his own magazine – *Ally Sloper's Half-Holiday*. Its title referred to the growing practice of giving workers a holiday on Saturday afternoons, which was helping to create the modern 'week-end'. The masthead claimed that the new magazine was 'founded and conducted by Gilbert Dalziel', but the actual editor was once again Charles Ross, who in July 1884 handed the task of drawing Sloper to a young cartoonist named William Baxter.[23] To the delight of readers, Baxter not only populated his hugely detailed full-page cartoons with a vast Sloper 'family', but also gave his hero the freedom to mix with Royalty and public figures. Baxter's energy was infectious, and it was said that 'the phenomenal success of this journal dates from the time when he first took in hand the eccentric "Ally"'.[24] However, the hard-drinking Baxter was eventually sacked, and the job of drawing Sloper given to one of his deputies, William Thomas.[25]

Baxter's cartoon was by no means the only one about Bismarck's resignation. Already, on 29 March 1890, *Punch* had published John Tenniel's famous cartoon 'Dropping the Pilot', showing Bismarck being effectively dismissed by the Kaiser. This was a classic *Punch* 'big cut', as the political cartoon was known, from the fact that it was still produced by wood engraving.[26] Thomas' line block was a far more demotic image, depicting Sloper and Bismarck enjoying a drink and a pipe of tobacco, apparently after reading *Ally Sloper's Half-Holiday*, a copy of which lay on the ground. 'Tell yer vot it is mein freund,' Sloper says to Bismarck in fractured German, 'vot you vant is ein "half-holitag".' 'Ja, Ja, das ist gut,' Bismarck replies cheerfully. According to the original caption, Sloper also argued for the superiority of British gin over German beer, and 'got the best of it, both in the quality of discourse and the quantity of refreshment'.

In stark contrast to Gillray and Morgan, Thomas had produced a political cartoon for the mass market. Instead of relying on the inside knowledge of a small group of politically aware metropolitan readers, process engraving allowed him to address a readership of millions who might know nothing of individual politicians, beyond the fact of their existence. It marked the decline of the comic journalism of the past half-century, and of most titles except *Punch*. By 1890 *Fun* was losing money, largely through competition from cheap magazines using 'process', and the Dalziel business was in trouble. Two years earlier Gilbert Dalziel had bought his father's interest in *Judy* for £8000, but soon he was forced to sell it on.[27]

Ally Sloper might be a drunkard and a trickster, but he was a comic anti-hero with whom readers could identify. *Ally Sloper's Half-Holiday* responded to this demand by creating the first comic fan club – 'Sloper's Club' – and was soon bombarding his fans with all kinds of merchandising, including Ally Sloper's Relish Sauce, Ally Sloper paperweights, mugs, doorstops, walking sticks, tie pins and match cases. Ally Sloper also appeared on the stage, and in 1896 it was claimed that 'he stars in a full fifty per cent of our pantomimes'. In 1898 he featured in two of the first British films, with another two following in 1900.[28] Baxter and Thomas – who spent 30 years drawing Sloper – had succeeded in creating a comic hero as well known as any of the political figures he encountered.

Punch and its imitators continued to decline under the onslaught of cheap process work. *Fun* underwent a sharp drop in size, before being incorporated into *Sketchy Bits* in 1901, and *Judy* managed to limp along only until 1907, but the next image shows where some of their energy went. This cartoon was drawn by the 36-year-old William Kerridge Haselden, for publication in the *Daily Mirror* of 2 July 1909 (Fig. 2.4). With the caption 'Votes and Violence', its subject was the rise of suffragette militancy. 'I've smashed windows,' says the suffragette, 'I've smacked an Inspector's face ... I've knocked his cap off ... I've used the whip ... I've tried to pull a policeman off his horse ... And yet they won't give me the vote!'

Haselden was a self-taught cartoonist, who in 1902 had joined the staff of *Sovereign* magazine to draw theatrical sketches and 'a sort of political cartoon of my own invention'. Unfortunately *Sovereign* folded soon after, leaving him looking for a job.[29] But in November 1903 Alfred Harmsworth, the newspaper and magazine publisher, launched a penny daily paper written by women for women, called the *Daily Mirror*, and in the following month Haselden wrote to him asking for a job.[30]

In fact the *Daily Mirror* had spectacularly failed to find a market, but Haselden's letter was referred to Arkas Sapt, one of Harmsworth's employees. Sapt had been charged with rescuing the paper, in company with its new editor, Hamilton Fyfe, and the journalist Kennedy Jones. The three men were far from confident that they could find a new readership, with Jones later admitting that 'we were in two minds about the future of the paper'.[31] But Sapt nevertheless decided to offer Haselden employment at £5 a week, telling him 'that they wanted a cartoonist, and they would boom me like a second F.C. Gould'.[32]

Figure 2.4 William Kerridge Haselden, 'Votes and Violence'

   The comparison was a significant one. Francis Carruthers Gould was one of the best known British political cartoonists. He had joined the *Pall Mall Gazette* in 1887 as the first political cartoonist ever to work on the staff of a British daily paper,[33] and within ten years it was noted that London evening papers like the *Pall Mall Gazette* were now the focus of political cartooning. Magazine publishers were abandoning political controversy in their search for increased circulation, noted one writer in 1897, and 'a powerful cartoonist or satirist on the staff would be regarded as an element of danger':

> Readers who want topical caricatures or strong cartoons seek them in the *Star* or the *Evening News* and not in *Punch, Judy, Fun, Moonshine*, and the other minor satellites. In other words, the papers that profess to be humorous are abandoning a political propaganda for which they feel themselves unfitted and the ordinary newspapers ... are running it for all that it is worth.[34]

Sapt was keen to introduce this form of lively political debate into the *Daily Mirror*, by making Haselden the first staff political cartoonist on a national morning paper. He started work in the middle of January 1904, drawing single-frame political cartoons similar to those he had produced for the *Sovereign*.[35]

The *Daily Mirror* was desperately searching for a new market, and Haselden was an important part of that campaign. At the end of January 1904 the paper made a dramatic bid for circulation by dropping its price to ½d. and committing itself to reporting photographically 'news which was formerly only told in words'.[36] Over £150,000 was spent on making the change,[37] but the circulation immediately jumped, and by the end of the year had reached 240,000 copies a day.[38] The *Daily Mirror* was changing around Haselden, noted one contemporary, as it 'gradually ... became less of a political and more of a family newspaper'.[39] Haselden not only had problems in finding a style suited to the new mass-circulation paper, but also acknowledged that at first he 'didn't know what sort of ideas were wanted'.[40]

Each morning Haselden read through the other London papers, and tried to pick up any new social fad or phenomenon 'before it has been thrashed to death in the drawing-rooms of Suburbia'.[41] He then drew his cartoon in black ink on art board, working to meet the deadline for the first edition of the *Daily Mirror*, at 9 p.m.[42] The printers turned his drawings into process blocks, which were then set in a page of type, from which a mould was made to cast semicircular printing plates for the paper's presses. By 1909, when this cartoon was published, the *Daily Mirror* employed huge Goss machines, carrying four separate reels of paper and capable of printing 25,000 copies per hour of a 16-page paper.[43] By July 1910 the paper's circulation had passed 650,000, rising to 750,000 by July 1911, 850,000 by July 1913 and 1,050,000 on the eve of war in July 1914.[44]

The *Daily Mirror* was now the largest selling daily paper in the world, delivering Haselden's cartoons to the biggest regular readership ever addressed by a cartoonist. However, the paper was increasingly directed at the home market, for the simple reason that most purchasing decisions were made by women, and domestic readership was enormously valuable in attracting consumer advertising. Papers like the *Daily Mirror* thus adopted a 'brightness and a chatty freshness' that was thought to appeal to women, and which proved very successful in encouraging domestic readership.[45] By July 1914 the *Daily Mirror* was proclaiming itself 'essentially a home journal': it was more often delivered by a newsagent than bought casually, had a high proportion of female readers, and enjoyed 'a longer life in the home than any of the heavier text papers'.[46]

As the *Daily Mirror* moved into this new market, Haselden had to adapt his cartoons to suit. 'He was not strong in ideas,' Hamilton Fyfe recalled, 'but he was, for a cartoonist, very quick at picking them up.'[47] He gradually abandoned the single-frame political cartoon, and in 1907 committed himself to the multi-frame format, usually of six images, which would be his trademark for the next three decades. The theme, as Haselden expressed it, was the 'little passing topics of the day'.[48] At the end of 1907 the first collection of these social-comment cartoons appeared in book form, under the title *Daily Mirror Reflections*. It was the first volume of non-political cartoons ever issued by a British newspaper, and the series continued for decades.

This cartoon by Haselden may thus appear easy to understand, but it is in fact rich in context. It may be a political cartoon of sorts, but it has been tempered by the commercial needs of the paper. Although Haselden began as a political cartoonist, he was forced to adapt as the *Daily Mirror* changed its position in the market, until eventually, as one commentator noted, he became 'essentially the "family cartoonist", content to illustrate the small humours, vexations, and incidents of ordinary life'.[49] There were thus tensions between the subject of this cartoon, the newspaper in which it appeared, and the naturally conservative opinions of the cartoonist. Haselden was free to comment on the campaign for female suffrage, but the comment had to be light, and of a type that would not alienate the paper's many female readers.

Haselden had been drawing suffragettes since April 1906, when he first showed them storming the House of Commons with axes and a gun. They remained, as one of his cartoons from March 1907 put it, 'A Significant Shadow' in British politics, and became one of Haselden's stock characters. In February 1908 he even parodied this with a drawing entitled 'A cartoonist's complete outfit', which naturally included 'One suffragette' among the useful stereotypes.[50] This drawing is perhaps not typical of his cartoons on this subject, many of which showed the suffragettes as unattractive and unfeminine,[51] but, perhaps with an eye to the paper's large female readership, he was always careful not to demonize them as individuals.

Not offending the reader was vital for cartoonists on mass-circulation publications – even the London evening papers, which remained the focus of political cartooning. The next image to be discussed, by the 34-year-old David Low, appeared in *The Star* on 18 August 1925. On the previous day the cricketer Jack Hobbs had finally scored his 126th century in first-class cricket, equalling the record set by W.G. Grace.[52] Press attention had been focused on Hobbs all season, and Low's cartoon showed him installed in the 'GALLERY OF THE MOST IMPORTANT HISTORICAL CELEBRITIES'. As the plinths revealed, the smaller figures alongside him were Adam, Julius Caesar, Charlie Chaplin, Muhammad, Christopher Columbus and David Lloyd George, the Prime Minister. They all looked up to the towering Hobbs with surprise and admiration, the caption noting that compared to the others he was absolutely 'IT' – the centre of attention and the only topic of conversation.[53]

The circulation of the Liberal *Star* was around 700,000 copies a day, a significant number, although some 500,000 less than the Conservative *Evening News*, the market leader among London evening papers.[54] Low considered his cartoon 'a piece of mere facetiousness, meaning nothing', but the editor, Wilson Pope, immediately grasped its popular appeal and printed it twice the usual size.[55] It was a shrewd move, for, according to *The Star* a few days later, Low's cartoon 'delighted the cricketing world and brought us a large number of letters eulogising both the draughtsmanship and genius of the cartoonist'. According to one regular reader, it was 'one of the cleverest and most humorous cartoons that Low has ever produced': 'The selection of the minor statues is almost a stroke of genius.'[56]

Low was surprised by the exaggerated praise for his cartoon, but he was even more surprised when an 'indignantly worded protest' arrived from the Ahmadiyya Muslim Mission in London.[57] The Mission was just laying the foundations for the capital's first mosque, and it took offence at the inclusion of Muhammad as one of the minor statues, looking admiringly up at Hobbs.[58] Wilson Pope wrote to the Mission, and 'expressed his regrets at the unintentional offence', but this did nothing to reduce the sense of outrage. Far from being satisfied with the response, it seems that the London Ahmadiyya Mission got in contact with other branches of the Mission in India, where news of the cartoon soon began to circulate.

By mid-October 1925 a protest campaign against Low's cartoon was underway. In Calcutta, noted the correspondent of the London *Morning Post*, the complaints now extended to the representation of Adam, who is also a Muslim prophet:

> An Urdu poster has been widely circulated throughout the city, calling upon Moslems to give unmistakeable proof of their love of Islam by asking the Government of India to compel the British Government to submit 'the ill-mannered editor of the newspaper to such an ear-twisting that it may be an object-lesson to other newspapers'.[59]

Rapid international communication had brought a reaction that Low and Pope could hardly have expected. On 16 October 1925 meetings were held in Calcutta mosques, resulting in resolutions and prayers 'calling upon the Government of India to make immediate representations to the British Government regarding measures to be adopted to prevent the recurrence of such outrageous conduct'.[60]

There was thus a huge contrast between the different responses to Low's cartoon, none of which can be read from the image alone. For the majority of *The Star*'s London readers, it was simply a mild joke about the achievements of Hobbs and the nature of fame, featuring a miscellaneous group of historical figures. But to some Muslims it was a deliberate insult. As the *Morning Post* correspondent wrote from Calcutta:

> There is no doubt whatever that, quite unwittingly, the cartoon has committed a serious offence which, had it taken place in this country, would almost certainly have led to bloodshed. What was obviously intended as a harmless joke has convulsed many Moslems to speechless rage, for while there is some laxity among them as regards the religious law against the making of pictures, no one has ever dared to attempt to depict Mohamed. When a picture of the Prophet appears in a cartoon, no explanation will suffice; it is an insult.[61]

Low felt that such an offence was unavoidable on a large-circulation paper, as there were 'always too many toes to tread on'. But he had already been warned about including religious elements in cartoons. On 23 June 1920 *The Star* had carried one of his cartoons showing the heavily armed leaders of the wartime Allies, including Lloyd George, heading for the Spa Conference dressed as angels. Entitled 'The Peripatetic Angels of Peace', it led to a complaint from *The Star*'s proprietor, Henry Cadbury, that this borrowing of Christian religious symbolism was in 'bad taste', and that 'as a general proposition ... laughter was unseemly when associated with religious matters (i.e. the accoutrements of angels)'.[62]

Low seems to have heeded the earlier warning, but only in relation to Christianity. When he chose a religious figure for his cartoon about Hobbs he thus did not select Jesus, who would have been much more familiar to readers of *The Star*, but whose inclusion might have produced accusations of blasphemy. Instead he chose Muhammad, misunderstanding the depth of offence he would cause to a small minority of *The Star*'s readers, and how that offence might spread abroad. 'The whole incident showed how easily a thoughtless cartoonist can get into trouble', Low wrote later. 'I had never thought seriously about Mahomet. How foolish of me. I was ashamed – not of drawing Mahomet in a cartoon, but of drawing him in a silly cartoon.'[63]

Similar tensions are apparent in the final cartoon, from *The Independent* of 14 December 2001 and showing the Palestinian leader Yasser Arafat (Fig. 2.5). Drawn by the 44-year-old Dave Brown, it is recognizably in the same tradition as Gillray's etching of Lady Hamilton, created 200 years before. It followed the declaration of the Israeli government, under the premiership of Ariel Sharon, that Arafat had been 'directly responsible' for the ambush of an Israeli bus two days before. Israeli tanks had been moved into the Palestinian city of Ramallah, where Arafat had his headquarters, and one was parked with its gun pointed at the building. 'SHARON AIMS GUNS AT ARAFAT AS TIES ARE SEVERED' read a headline in the issue of *The Independent* which carried Brown's cartoon: 'PALESTINIAN LEADER ENFEEBLED AND BESIEGED IN HIS HEADQUARTERS'.[64] The Israeli government said it had no intention of killing Arafat, but a helicopter had already fired a missile into a building only 20 yards from his office.

Brown's response to these events was a brilliant pastiche of William Holman Hunt's 1855 painting *The Scapegoat*. That had depicted the Jewish sacrificial goat driven into the wilderness on the Festival of the Day of Atonement, and Brown created a powerful political metaphor by transforming the goat carrying the sins of the Jewish community into Yasser Arafat, firmly tethered in the desert. He retained many details of the original painting, such as the horns of the dead ibex in the background, but he transformed the scarlet woollen cloth around the goat's horns into Arafat's keffiyeh, and the goat's skull in the sand into the skeleton of a long-dead dove of peace.

The cartoon was one of a series by Dave Brown, drawing attention to the inconsistencies and inequalities in the relationship between Israel and the Palestinians. It was created for the educated readership of a London broadsheet newspaper, whose circulation was around 200,000 copies a day – a fraction of the circulations of *The Times*, the *Telegraph* and the *Guardian*, its rivals among 'quality' newspapers, which together sold almost two million copies. Brown's cartoons reflected the policy of *The Independent* itself, and yet, as the second version reveals, his cartoon of 14 December 2001 had in fact begun as a far more aggressive image (Fig. 2.6).

The original artwork, in pen and acrylic on card, had the breadth of Holman Hunt's original canvas, and included a devastating caricature of Ariel Sharon in a very phallic tank costume, about to assault Arafat. This was the version delivered to *The Independent*, but its editorial staff decided to modify it without consulting the cartoonist, creating the smaller published version. This may have been designed to forestall criticism, but it was not a concerted attempt to muzzle the cartoonist, for on 27 January 2003 Brown included

Figure 2.5 Dave Brown, 'The Scapegoat'

Figure 2.6 Dave Brown, 'The Scapegoat' (original version)

Sharon in an even more devastating cartoon pastiche for *The Independent*. Drawn in the wake of an Israeli attack on Gaza City, in the run-up to the elections for the Knesset, this was a dark comment on what Brown considered a 'pretty perverse' form of electioneering.[65] As attack helicopters broadcast political slogans, the cartoon showed Sharon biting the head off a child, and saying 'What's wrong ... you never seen a politician kissing babies before?'

This was a tangle of references borrowed from popular culture, classical mythology and fine art. The helicopters might come from the film *Apocalypse Now*, but, as Brown acknowledged in the margin, the main composition was taken from Francisco Goya's 1819–23 painting of Saturn eating one of his own children, fearful of a prophecy that they would eventually overthrow him. The use of this image, taken from Goya's series of 'Black Paintings', to suggest that Sharon was killing those he should protect, made the design especially powerful. As Brown explained, 'my first idea was of Sharon puckering up to a child, revealing missile-like fangs':

> Then my thoughts progressed from biting to eating children, and immediately Goya's painting Saturn Devouring One of His Sons came to mind. ... By borrowing the image, I hoped to benefit from its associations; those who knew the classical myth of the Titan driven, by his fear of being supplanted by his children, to the insanity of devouring them, might draw some parallels.[66]

The cartoon may have been intelligible to *The Independent*'s regular readers, but to those unfamiliar with Goya's painting, and the British cartoon tradition, it carried a quite different meaning, and protests poured in. To the Israeli Embassy Brown's design resembled the cartoons in some Arab newspapers, based on the ancient 'blood-libel' that Jews slaughtered Christian children and used the blood in their rituals. Condemning it as a cartoon 'which would not have looked out of place in *Der Sturmer*' – the viciously anti-Semitic Nazi magazine[67] – the Embassy took its protest to the Press Complaints Commission (PCC). The PCC, however, decided that the cartoon contained 'nothing ... that referred to Mr Sharon's religion', and was valid political comment.[68]

With regard to the question of context and interpretation, the PCC added that it was 'unreasonable to expect editors to take into account all possible interpretations of material they intend to publish'. This was especially appropriate for an age in which political cartoons were easily published and republished, and could be distributed around the world within hours. There may be visual similarities between the work of Gillray and Brown, but whereas Gillray addressed a small coherent readership largely within London, Brown's cartoons might easily travel far outside their original context, to gather new and unintended meanings as they crossed national and cultural boundaries. This was indeed the case with his pastiche of Goya, which within months had been copied onto a placard in India, and was carried by radical Muslims demonstrating against Israel.[69]

Paraded in front of an audience unfamiliar with the layers of referential meaning in British political cartooning, Brown's cartoon appeared in an entirely new light. This is further confirmation of the need for academics to locate such visual images within a precise context, rather than treating them simply as illustrations. Satirical prints and cartoons can be useful illustrations that catch the eye of the reader, but they are far more valuable as evidence of an important set of dynamic social and political relationships. To use them as evidence academics need to be aware of the contexts in which they sat: the changing technology of image creation, manufacture and distribution; the markets and readerships they served; and the graphic artists, editors and publishers who produced them. The effort of understanding will be amply repaid by the rich evidence they provide.

Notes

1   Dorothy George, *Catalogue of Political and Personal Satires Preserved in ... the British Museum: Vol. VIII, 1801–1810* (London: British Museum, 1978), 37–8.

2   Walpole to Mary Berry, 11 September 1791, in *Horace Walpole's Correspondence: Volume Eleven*, ed. W.S. Lewis and A. Dayle Wallace (New Haven, CT: Yale University Press, 1944), 349.

3   Patricia Jaffé, *Lady Hamilton in Relation to the Art of her Time* (London: Arts Council of Great Britain, 1972), 68–73.

4   Brian Fothergill, *Sir William Hamilton: Envoy Extraordinary* (London: Faber & Faber, 1969), 397.

5   Draper Hill, Introduction to *James Gillray 1756–1815: Drawings and Caricatures* (London: Arts Council, 1967), 6–8.

6   Hill, Introduction, 8.

7   George, *Catalogue*, 28–30. 'Political Dreamings!' appeared in November 1801.

8   Diana Donald, *The Age of Caricature: Satirical Prints in the Reign of George III* (London: Yale University Press, 1996), 184.

9   Vic Gatrell, *City of Laughter: Sex and Satire in Eighteenth-Century London* (London: Atlantic, 2006), 235.

10 'Where is Britannia?', *The Tomahawk*, 8 June 1867, 51.

11 E.E.P. Tisdall, *Queen Victoria's Private Life 1837–1901* (London: Jarrolds, 1961), 87–91.

12 'Graphotype', *The Times*, 11 September 1869, 8.

13 Arthur William à Beckett, *The à Beckett's of* Punch: *Memories of Father and Sons* (New York: E.P. Dutton, 1903), 164.

14 Alvar Ellegård, *The Readership of the Periodical Press in Mid-Victorian Britain* (Gothenburg, Sweden: University of Gothenburg, 1957), 35–8.

15 Thomas Milton Kemnitz, 'Matt Morgan of "Tomahawk" and English Cartooning, 1867–1870', *Victorian Studies* 19, no. 1 (1975): 9. Kemnitz surveyed 'cartoons of opinion' during 1868–9.

16 A.W. à Beckett, *The á Beckett's of* Punch, 159, 161.

17 Elizabeth Longford, *Victoria R.I.* (London: Weidenfeld & Nicolson, 1964), 330. A sale of 30,000 copies was claimed by the magazine in 'Important Notice', 19 October 1867, p. 240.

18 'John Brown, or the Fortunes of a Gillie', *Liverpool Mercury*, 12 October 1867, 5.

19 'Owl-la Podrida', *The Owl*, 6 April 1883, 6.

20 James Thorpe, 'Mr. Fletcher-Thomas: Memories of "Ally Sloper"', *The Times*, 26 March 1938, 17; Peter Bailey, 'Ally Sloper's Half-Holiday: Comic Art in the 1880s', *History Workshop Journal* 16, no. 1 (1983): 9–10.

21 His first appearance was in *Judy*, 14 August 1867, 199, 'Some of the Mysteries of Loan and Discount'.

22   It is not true, as often stated, that Sloper's name was derived from his supposed habit of ducking down back alleys to escape the rent collector. An earlier incarnation was ' 'Arry Sloper', a character invented by Charles Ross for a story published in *Reynolds' Miscellany* in 1862: John Adcock 'The Creator of Ally Sloper', http://yesterdays-papers.blogspot.com/search/label/Judy (accessed 26 December 2006).

23   'Mr. Gilbert Dalziel', *The Times*, 14 May 1930, 21; Bailey, 'Ally Sloper's Half-Holiday', 7–8; Alan Clark, 'Charles Henry Ross', *Dictionary of British Comic Artists, Writers and Editors* (London: British Library, 1998), 149.

24   'The Late W.G. Baxter', *The Graphic*, 4 August 1888, 114.

25   Thorpe, 'Mr. Fletcher-Thomas'; Simon Heneage, 'William Fletcher Thomas', in Mark Bryant and Simon Heneage, *Dictionary of British Cartoonists and Caricaturists 1730–1980* (Aldershot, UK: Scolar Press, 1994), 220; Clark, 'William Fletcher Thomas', *Dictionary of British Comic Artists*, 162–3.

26   Rodney Engen, *Sir John Tenniel: Alice's White Knight* (Aldershot, UK: Scolar Press, 1991), 140–2.

27   'High Court of Justice', *The Times*, 2 November 1893, 10; Frederic Boase, 'Charles Henry Ross', *Modern English Biography*, vol. 4 (London: Frank Cass, [1921] 1965), cols. 497–8.

28   Roger Sabin, 'Ally Sloper: The First Comics Superstar?', *Image and Narrative: Online Magazine of the Visual Narrative* no. 7 (October 2003), www.imageandnarrative.be/graphicnovel/rogersabin.htm.

29   British Cartoon Archive, University of Kent (BCA), W.K. Haselden, 'Answers to Questions Put to Me by Richard Jennings', undated typescript copy, 2.

30   Adrian Margaux, 'Our Social Satirist: W.K. Haselden and his Work', *Pearson's Magazine*, August 1917, 152–3.

31   Kennedy Jones, *Fleet Street and Downing Street* (London: Hutchinson, 1920), 233.

32   BCA, Haselden, 'Answers to Questions', 2.

33   Mark Bryant, 'Sir Francis Carruthers Gould', *Dictionary of Twentieth-Century British Cartoonists and Caricaturists* (Aldershot, UK: Ashgate, 2000), 95.

34   'Our Political Comic Papers', *Reynolds's Newspaper*, 28 February 1897, 2 (letter signed 'Gracchus').

35   Haselden's first published cartoon was 'The Bear with the Open Door', *Daily Mirror*, 11 January 1904, 7. The online catalogue of the British Cartoon Archive contains 5700 of Haselden's *Daily Mirror* cartoons, taken from the original drawings.

36   'New Epoch in Journalism', *Daily Illustrated Mirror*, 29 January 1904, 3.

37   Daily Mirror, *The Romance of the Daily Mirror 1903–1924* (London: Daily Mirror Newspapers, 1925), 4.

38   Henry Simonis, 'The Street of Ink: The Daily Mirror', *Newspaper World*, 20 January 1917, 16.

39   Margaux, 'Our Social Satirist', 153.

40   BCA, Haselden, 'Answers to Questions', 3.

41   'The Satire of W.K. Haselden', *Strand Magazine*, November 1908, 521.

42   British Library, Northcliffe Papers, Add.Mss.62234, ff.27–8, Alex Kenealy to Northcliffe, 27 March 1909.

43   George A. Isaacs, *The Story of the Newspaper Printing Press* (London: Co-operative Printing Society, 1931), 85–7.

44   *Newspaper Press Directory 1911* (London: Mitchell, 1911), 434; *Daily Mirror* advertisement:

Advertising World, October 1911, 467; *Daily Mirror* advertisement: Circulation Manager, February 1914, 17; *Daily Mirror* advertisement: St Bride Printing Library, London, *The Daily Mirror and its Circulation: August 1914* (London: Pictorial Newspaper Company, 1914), 13.

45   Mary Frances Billington, 'The Woman as Reader: Feminine Influences on Newspapers', *Sell's World's Press 1915* (London: Sell's, 1915), 46.

46   St Bride Printing Library, *The Daily Mirror and its Circulation*, 7, 20, 23.

47   Henry Hamilton Fyfe, *My Seven Selves* (London: Allen & Unwin, 1935), 98.

48   BCA, Haselden, 'Answers to Questions', 3.

49   'Mr. W. K. Haselden: Kindly Cartoonist', *The Times*, 29 December 1953, 9.

50   BCA, WH5524, Haselden cartoon of 15 February 1908.

51   Judith S. Kindred, 'The Portrayal of Women in the Cartoons of William Kerridge Haselden 1906–1930' (MA thesis, University of Kent, 1997), 34.

52   'Cricket / Hobbs 101 / The Champion's Record Equalled', *The Times*, 18 August 1925, 6.

53   This cartoon is reprinted in Colin Seymour-Ure and Jim Schoff, *David Low* (London: Secker & Warburg, 1985), 63 and is Image LSE7298 in the catalogue of over 9500 of Low's cartoons on the British Cartoon Archive website at www.cartoons.ac.uk.

54   A.P. Wadsworth, 'Newspaper Circulations, 1800–1954', *Transactions of the Manchester Statistical Society*, Session 1954–5 (Manchester, 1955), 37.

55   David Low, *Low's Autobiography* (London: Michael Joseph, 1956), 123.

56   BCA, photocopy of David Low's cuttings book, undated cutting from *The Star*.

57   Low, *Autobiography*, 123.

58   'New Moslem Mosque Near Wimbledon', *The Times*, 29 September 1925, 17.

59   BCA, photocopy of David Low's cuttings book, cuttings from the *Melbourne Herald*, *c.*12 October 1925, and *Morning Post*, 3 November 1925.

60   BCA, photocopy of David Low's cuttings book, cutting from the *Morning Post*, 3 November 1925.

61   BCA, *Morning Post*, 3 November 1925.

62   Low, *Autobiography*, 122. Low recalled the cartoon rather differently, but it seems to have been this one – LSE6324 in the BCA catalogue.

63   Low, *Autobiography*, 124.

64   Phil Reeves, 'Sharon Aims Guns at Arafat', *The Independent*, 14 December 2001, 17.

65   Ed Caesar, 'Drawn into the Row', *The Independent*, 6 February 2006, 10.

66   Dave Brown, 'The Cartoonist Writes', *The Independent*, 31 January 2003, 6.

67   'Jewish Stereotype' (letter from Shuli Davidovich), *The Independent*, 28 January 2003, 17.

68   'Press Watchdog Says *Independent* Cartoon of Israeli PM was not Anti-Semitic', *The Independent*, 21 May 2003, 6.

69   http://backspin.typepad.com/backspin/2003/11/evolution_of_an.html (accessed June 2007).

3 'Impressed by nature's hand': photography and authorship

by Douglas R. Nickel

The legal term *res ipsa loquitur*, commonly translated as 'the thing speaks for itself', refers to a claim that is manifestly self-evident, one so blatantly apparent that elaboration of the matter would be pointless. Photography has always seemed implicated in the conceptual equivalent of this doctrine.[1] Paintings, sketches and other hand-crafted varieties of pictorial representation may well depict a subject found in the real world – the likeness of a person recognizable in a portrait, say – but these media can with equal ease represent something drawn wholly from the artist's imagination, and in the end there is nothing in the image that guarantees the source of the subject resides in one location or the other. Verbal and written texts too share this relation to their subjects. The camera photograph, on the other hand, requires its subject to exist physically and materially, to have some reality independent of the photographer's attention to it, before the thing or scene can be rendered by light and chemistry as an image.[2] Indeed, this has been deemed the necessary condition of the process. Most would agree that even subjects that are not visible to the human eye and observable only as artefacts of the act of photography (blurs, for instance, or moving objects frozen by stop action techniques) correspond to *something* self-sufficiently real that made the phenomenon on the photographic plate possible. Thus, photography's ontology – its nature, its essential relationship to the phenomenal world – appears at first glance to be categorically different from that of representations generated without the use of lenses and light-sensitive materials. And yet we know that the photograph is not merely the result of a process or an apparatus but also, always, the product of intention, selection, editing, chance, desire, convention and ideology: a cultural object, in other words, the outcome of human will and interest. Its ability to communicate ideas or feelings depends in large measure upon the context in which it is found and the capacity of its viewer to decode meanings implicit in it. Considered as a kind of visual evidence, then, the photograph stands out as a special, and especially problematic, forensic object, one that may seem to speak for itself even as it is in fact speaking for its makers. To understand what kind of claims the photograph makes, we must first appreciate the stakes of asserting its essential differences over its essential similarities to other kinds of pictures.

Much confusion about the photograph arises from the vexed issue of authorship. Even though photographs display properties generic to all pictures, they appear in a way the creation of physical laws and automatic operations over which their human maker can exercise but some finite degree of control. The robotic camera in the bank lobby makes an image that looks identical to the one that would result had a living photographer stood in the same place. In this sense the process itself becomes the author of the image. Granted, a technician must set up and programme the mechanism to do its job – much as a scientist will set up an experiment with possible outcomes in mind – but the results that matter happen only when the technician or scientist removes himself from its performance and let the system run. In a way, all photographs are haunted by this implicit sovereignty: the machine may be guided by its operator, but it can also make pictures without direct management or control. This seems to impart a fundamental neutrality upon the result.

Even within its four edges, the photograph retains the means to record – reflexively, as it were – what may be unnoticed by the photographer at the time the image is exposed. Here one need only remember Antonioni's 1966 film *Blowup*, in which the protagonist later realizes that, while photographing lovers in a park, he may have inadvertently recorded a murder taking place in the background of his composition. Throughout its history, the medium's authority has flowed from a shared sense of the photograph's imperviousness to internal manipulation, from an awareness of the mechanical way the lens must obey the laws of optics to arrange its image on a flat surface and the inexorable way chemicals must register that optical image. Photography appears distinct from other media in this respect. The smallest unit of manual addition or subtraction in a drawing is the line; in a painting, the brushstroke; in language it is the letter, character or phoneme. Drawings and paintings are built cumulatively of lines, tones and details; texts are built cumulatively of letters, words, sentences, paragraphs and so on. In contrast, the photograph itself is the smallest functional unit of photography: the lens typically projects a visually cohesive perspectival image bounded by a frame, and pictorial components within that frame can only with extreme difficulty be altered and still remain integrated.[3] The lens often gives the impression of enveloping the photographic image in an atmosphere of continuous tone. When the photograph is made, then, it is not made up of parts, individually considered, applied, adjusted and then modified with other parts. It is made, we know, all at once.

The necessary presence of a concrete referent is therefore regarded as the first degree of the medium's relationship to its subject. Not only is the subject required, but also prevailing opinion holds that the subject somehow *causes* the image. In the 1860s, the American philosopher Charles Sander Peirce developed a general theory of signs, wherein three of the signs he discusses are based on their relationship to the objects they signify. He designates these the 'icon', the 'index' and the 'symbol'. As Peirce explains, an icon represents through likeness, through recognizable shared qualities, as when a drawing or caricature resembles its subject. A symbol represents not through resemblance, but through habit, convention or consent. For example, in English the word 'dog' is a symbol that stands for a certain kind of four-legged mammal, while 'chien' stands for the same in French. The sign here has an arbitrary relation to its subject, as neither word derives its form from actual dogs but purely from the exigencies of discrete languages. An index, finally, is a sign caused by its referent, what Peirce calls a 'correspondence in fact' between object and sign.[4] Resemblance is not required, only modification through connection or cause-and-effect: the smoke rising from a chimney will indicate a fire in the hearth below, for instance, just as a footprint in the snow becomes an index of the boot that caused it. Among the several ways Peirce came to define the term we find:

> [The index] is a real thing or fact which is a sign of its object by virtue of being connected with it as a matter of fact and by also forcibly intruding upon the mind, quite regardless of its being interpreted as a sign. It may simply serve to identify its object and assure us of its existence and presence. But very often the nature of the factual connexion of the index with its object is such as to excite in consciousness an image of some features of the object, and in that way affords evidence from which positive assurance as to truth of fact may be drawn. A photograph, for example, not only excites an image, has an appearance, but, owing to its optical connexion with the object, is evidence that that appearance corresponds to a reality.[5]

Significantly, Peirce is not arguing it is indexicality that makes the photograph correspond in appearance to its subject, but rather that the index can, through inference, call to consciousness some awareness of the thing that caused it. His system was never intended as a sorting mechanism for placing signs into categories, but merely a logical description of qualities that signs possess. As these qualities are not mutually exclusive, Peirce can describe the photograph, as he does here, as having an appearance that resembles its subject (iconicity) *and* a perceived connection to its source (indexicality), the icon and index conjoined. Most of the heavy lifting the photograph does as visual evidence would seem to depend upon the way its subject strikes the viewer as looking very like the thing represented: a photograph caused by its object that did not look like that object would not be of much use to most of us. However, current theoretical writing on photography all too often takes the very qualities Peirce was trying to put into relation, one to another, and conflates them, figuring the index as not about causal relationship *tout court* but rather any one-to-one correspondence of subject and image. 'Index' is used interchangeably with words like 'trace', 'impression' or 'imprint', locutions that might make an apt analogy if characterizing the indexical properties of, say, a stencil or fingerprint, yet are plainly unhelpful when describing what is indexical about chimney smoke, weathervanes or a photograph.[6] For, as Joel Snyder has maintained, in the photographic process nothing is ever actually traced or impressed. Light, whether resolved by a lens into an image or passing through a stable matrix (a negative, for example), alters the structure of photosensitive particles at a molecular level and, either directly or by means of subsequent chemical applications, causes some of these particles to change composition, to lighten or darken. An image may or may not result. The trope of 'impression' or 'trace' logically implies the motion of one object marking another through direct physical contact. This aspect of the metaphor is misleading to an understanding of photography, for whatever physically causes the photograph to happen, it is clearly not this.[7]

The second degree of photographic representation, then, concerns its analogic association with its object – the specific manner in which the image can be seen as relational to what was before the camera. When viewing a photograph – and here we are considering only ordinary photographs with recognizable subjects – we typically regard its surface as more or less transparent, as if we were looking *through* to a subject within or beyond it, rather than *at* a flat ground of various tones or hues. This is nothing so simple as an outright confusion of subject and object, for we always retain some awareness that the depiction in front of us was excerpted out of the continuum of the visible world, that it was *framed*, and such framing invariably points back from transparency to a selecting intelligence, to intention.[8] Nonetheless, in practice we experience something effectively coded as (what we take to be) 'real' space and solidity, an effect that collapses subject and object. The critic Roland Barthes spoke suggestively of this effect. 'The Photograph belongs to that class of laminated objects whose two leaves cannot be separated without destroying them both,' he writes. 'Whatever it grants to vision and whatever its manner, a photograph is always invisible: it is not that we see.'[9] As with a landscape viewed through a windowpane, the image equivocates between a bounded field and flat surface and a three-dimensional realm that we envision has been mapped on to it. We can 'delaminated' the effect, in Barthes's terms, by moving our attention away from the imagined space and solidity of the representation and to the photograph as a surface and the boundaries of its frame. Here is where the seamless continuity of the illusion ends.

Barthes's eloquence on the matter notwithstanding, the transparency effect results not because of any innate physical property of photography, but because cameras came to be configured technologically to give results that were thought to approximate normal human vision (and that would approximate other pictures based on vision) and because photographers used them to that end.[10] This is to say, cameras were built and exploited to make a kind of picture that bore analogy to a window, just as Leon Battista Alberti, in *De Pictura* (1435), had proposed the representational picture as a kind of window. At the same time, object qualities of photographs – glossy, calendared or polished surfaces, for instance – were found to abet the sensation. Lenses, shutters, negative supports and print emulsions all developed in unison to arrive at the particular seamless brand of optical transparency and atmosphere we recognize as 'photographic'. Accordingly, these innovations and practices must be seen as historical, not natural: subject to the needs of individuals and society, in other words, and not a property of the medium. That illusionistic painting aspired to its own version of collapse and traded on its own techniques to create the impression of looking into the picture only underscores that photographic realism is a style, not a property.

Mimesis works through techniques that encourage the viewer to find the work's meaning in the meaning of the object represented rather than in the way it is represented. While the camera and its operators can make a certain kind of picture optically and chemically that appears transcriptive of a tangible subject or scene in the real world, it is not necessary that they do so whenever making a photograph. When a photographer stages a still-life arrangement of objects in a studio and renders that arrangement, the semantic balance gets tipped towards the selection of items and their meaning, to the arrangement, lighting, etc. – and proportionally away from framing. Here the possibility of unintended elements intervening is largely eliminated. While the camera ordinarily depicts the scene before it with a spacio-optical schema that approximates mathematical perspective and 'vision', this is a convention, not a requirement. The photographer can employ an extreme wide-angle lens or a completely uncorrected lens and produce images that are wildly distorted or so fuzzy as to make the subject illegible, and yet these would still be photographs. A photographer could copy something flat, like a map, and print it on a photographic paper whose surface is very similar to that of the original. The result would be more a facsimile of a map than the illusion of a map. To give a final, radical example, a photographer might make an exact copy of a piece of smooth white paper in flat light. If the processing is carried out properly, the result will be a smooth white sheet of photographic paper, simultaneously registering the original object and nothing. But it is still, nonetheless, a photograph. The precept illustrated here is that documenting real world subjects as spatial pictures is a *capacity* of the photograph, not an intrinsic property of it. Photography has other technological capacities, faculties not traditionally featured as an important part of its cultural identity. Photographs became evidence (and photography became photography) not because of inherent properties in the medium but because of the ways photography was used and the things that were said and thought about it.

Most arguments for photographic difference rely upon the medium's putative objectivity and contiguity with its referents to argue for a uniqueness, an inherent distinctiveness. But the point must be made – for it is the crucial point – that photographic authority does not reside in the photograph, but in the photographic viewer, as an effect or response to the viewing experience. This is not to argue that photography does not have capacities, or that we cannot identify and discuss the attributes of these capacities in specific photographs; it is only to insist that any attempt to define photography by way of innate or essential attributes will fail, because such definitions reduce a dynamic transaction into a static object. Photographic authority is not immanent in photographic objects, nor does it derive exclusively from the technological conditions of production for its objects; it is a psychological – and by no means historically inevitable – effect, a kind of investment in looking at photographs that involves a 'willing suspension of disbelief for the moment', as the poet Samuel Taylor Coleridge once put it.[11] Coleridge calls the effect 'poetic faith': the same mental operation that allows us to temporarily forget about actors and props on a stage and become absorbed in the action of a play allows us to forget about photographic surfaces, borders, even translation into black-and-white tones to become, to a remarkable degree, absorbed in a photograph. Modern theories of the photograph that stress the innate indexical or the self-generating nature of the medium (i.e. photography as essentially mechanical) invariably end in a stalemate with competing theories that present this nature as nothing but a mask behind which the photograph hides to impart its ideologically coded, manifestly constructed (i.e. essentially human) message.[12] Machine or human: which is the author? But one is forced to choose between these two contradictory and ultimately irresolvable positions only if wedded at the outset to defining the photograph ontologically. A phenomenological understanding recognizes that both the apparatus and the maker may be 'motivated', and that their mutual target is the viewer.[13]

Any statement couched as 'photography is ...' represents a theory. Many of the difficulties that have arisen around the photograph as a theoretical object can be resolved by attending to this basic distinction between what photography is and what photography is believed to be. The distinction is more than rhetorical, as many claims have been made on behalf of the latter as if they were accounting for the former. An example or two may clarify what is at stake. Suppose someone were to produce a highly detailed, handmade painting of their house, rendered in emulation of a photograph, and so very detailed and accurate in its portrayal of a lenticular image as to be impossible to distinguish from an actual photographic print, except upon very close examination. Now imagine this artefact is displayed in a museum, on a wall, in a frame, behind glass and a security barrier. Under these conditions the work looks exactly like a photograph – everyone believes it to be one. What would be its authority as evidence of the appearance of the artist's house? If we believe the object we are viewing is a photograph – an object with assumed ontological characteristics – we will likely accept its testimony about the house's appearance (existence, even) as credible: that is, until someone revealed its actual manner of derivation to us. The object will not have changed, only the kind of investment we make in the object, for, in a sense, it is we who create the object before us.[14] In this instance, it is not a real presence that matters, but an imagined presence.

Functionally, the photograph cannot claim authority; as an inanimate object, it literally cannot claim anything at all. But we can confer authority upon it. An anecdote published in an American trade journal in 1875 demonstrates the force of expectation in the pictorial transaction. A Mr Kardactz reports on his visit to China, and his encounter with a flourishing photographic studio in which one could find no sign of a camera.

> You could go in at any hour of the day, providing you were a Chinaman, and get a portrait executed in a very short time. ... The heathen Chinese had merely acquired a large collection of portrait negatives, and when a customer came, he took his measure mentally, looked through the stock, and chose the picture most like him. As all Chinese heads are pretty similar, and their pigtails much about the same length, it was never difficult, apparently, to make a match, for the public were quite content with what they got for their money.[15]

Presumably the reason patrons were content taking away someone else's portrait as their own is that, without prior experience of the photograph and no tuition in its workings or ontology, they had different expectations about how closely a photograph was supposed to match their appearance. As the scholar Yi Gu has noted, in the Chinese tradition painters often employed a 'face book' of standardized head structures and facial features, selecting from this repertoire matching elements that would be assembled into a new work. The result was not meant as a verbatim description of the sitter's individuality but the inscription of the client into a network of existing types. Customers looked for no more from the photograph, and so found what they sought.[16]

If we sort out the material conditions of its objecthood from the experience of its viewing, the photograph can be regarded as a proposition, one that satisfies a certain kind of expectation when approached with a certain hypothesis about its nature. By this account, it might be more accurate to characterize the photographic effect as one of immediacy rather than of transparency, inasmuch as this points to the psychology of the viewer rather than the physical make-up of the photograph. Inserting the viewer into the equation also inserts an important contingency, as photographic immediacy depends not simply upon an activating viewer, but one specifically trained in the set of assumptions required to make the reciprocity effect work. Different ontological assumptions, such as those we are now learning to bring to digital photography, jeopardize the effect, however similar to traditional photographs the images may appear.

So where did our expectations of the photograph come from? Examination of the historical record reveals that photography's invention as a technology was accompanied by a parallel invention, that of photography as an idea. This invention was discursive: it happened in culture, through language and actual photographic practice. The process of describing photography begins at the very moment in early 1839 when its discovery was publicly announced: two of its inventors, J.L.M. Daguerre and William Henry Fox Talbot, marshalled (respectively) the French government and British scientific academies around their innovations, in bids for support. Though Talbot had been experimenting since at least 1835 (and knew of Josiah Wedgwood and Humphry Davy's experiments of the 1790s) and Daguerre was at work on his process by 1829 (his partner Niépce had already succeeded in making permanent photographs two years early), the public first learned of the discovery in January 1839, when photography 'exploding suddenly into existence', as the photographer Nadar later put it.

> The appearance of the Daguerreotype ... was an event which ... could not fail to excite considerable emotion. ... It surpassed all possible expectations, undermining beliefs, sweeping theories away. ... It is impossible for us to imagine today the universal confusion that greeted this invention, so accustomed have we become to the fact of photography and so inured are we by now to its vulgarization.[17]

Even before many had actually seen a photograph, commentators found themselves striving to locate language to describe and account for something that to them seemed wholly unprecedented, 'a prodigy'. The inventors and those in their inner circles struggled with what to call the several processes: the name of the invention, after all, would point to salient features ('light', 'sun', 'nature', 'writing', etc.) to establish it as a concept.[18] How photography was first characterized by the inventors and their correspondents, how it was described by experts, officials, in the press and then in stories, plays, advertisements, court cases and popular speech – the overarching project was to put the unknown in relation to the known, to naturalize what was novel or unfamiliar about photography by reference to notions already commonplace to the audience of the time.

Photography was, by this means, constructed to be different – it was presented as like other things, but superior. This social construction presented the photograph as prototypically truthful, disinterested, self-generating, automatic, wondrous, even superhuman. The invention seemed to answer a need for representational means that could remove the unsteady human hand, the fallible human eye, and the subjective, distractible human mind from the task of reporting.[19] In 1847, a publisher in Boston inaugurated a periodical called the *Daguerreotype*. The journal did not discuss photography or reproduce photographs, but instead, according to its prospectus, it was to be a digest of literature and current affairs.

> The DAGUERREOTYPE is, as the name imports, designed to reflect a faithful image of what is going on in the Great Republic of Letters ... a picture in which the characteristic features will all be reflected, and of which, though the lights and shades may at time[s] be somewhat strongly marked, the general fidelity will be unquestionable.[20]

> Our name implies that we must portray every important feature. No partial or sectarian views must govern our choice, and even opinions from which we dissent must (when not of irreligious or immoral tendency) often find a page on our pages. A painting may omit a blemish, or adapt a feature of the artist's fancy, but a reflected image must be faithful to its prototype.[21]

The writer wishes to announce an editorial policy with his analogy, but, as Alan Trachtenberg notes, the figure works both ways: the daguerreotype is defined in turn as an object that bespoke utter objectivity, one lacking the ability to select yet betraying a fidelity to its subject through direct reflection that precludes human intervention or embellishment. By the 1850s, a 'daguerreotype' came to stand for any perfect copy of something.

About a year after photography's public announcement, Edgar Allen Poe celebrated the invention in an issue of *Alexander's Weekly Messenger*.

> All language must fall short of conveying any just idea of [it]. ... Perhaps, if we imagine the distinctness with which an object is reflected in a positively perfect mirror, we come as near reality as by any other means. For, in truth, the Daguerreotyped plate is infinitely (we use the term advisedly) is *infinitely* more accurate in its representation than any painting by human hands.[22]

Poe's image points to both the polished silver surface of the daguerreotype and the extant trope of the magic mirror, which features in Goethe's *Faust* (1801): in the witch's kitchen, Faust discovers a mirror that shows him the apparition of the ideally beautiful Helen, who becomes an object of the doctor's desire. The trope recurs in gothic stories and romantic poems in the decades before photography's announcement: even Talbot wrote one such ballad in 1830.[23] The Frenchman Charles-François Tiphaigne de la Roche published a story in 1760 about the mythical island of Giphantie. The hero of the tale escapes a hurricane to find himself confronted by a benevolent spirit – the island's prefect – who takes the form of a speaking cloud. Among the island's many wonders shown the hero is a room with a miraculous window-like frame; the protagonist is astonished to discover that its view of the departing storm is an illusion, fixed on the wall's surface. The prefect explains:

> That window, that vast horizon, those thick clouds, that raging sea, are all but a picture. ... Thou knowest that the rays of light, reflected from different bodies, make a picture and paint the bodies upon all polished surfaces, on the retina of the eye, for instance, on water, on glass. The elementary spirits have studied to fix these transient images: they have composed a most subtile matter, very viscous, and proper to harden and dry, by the help of which a picture is made in the twinkle of an eye ... [the] impression of the images is made the first instant they are received on the canvas, which is immediately carried away into some dark place; an hour after, the subtile matter dries, and you have a picture so much the more valuable, as it cannot be imitated by art nor damaged by time.[24]

This seeming prediction of photography some 80 years before the fact might well have receded into literary oblivion had it not been recovered by the Parisian photographers Mayer and Pierson and recounted in their 1862 book *La Photographie*.[25] Eighteenth-century fantasy literature and morality was thus interpolating into an emerging discourse about photography, where the new medium is endowed with mytho-poetic associations that propose it as unique, unprecedented, the modern technological fulfilment of what had previously been merely fantastic imaginings.

Language that proposed photography as 'infinitely more accurate in its representation than any painting by human hands' served to remove it from the world of artisanal craft and place in a modern realm of industrial mechanization.[26] Throughout the nineteenth century, the photographic image was continually proposed to be self- generating, or drawn by the sun, or by nature – not made by photographers. Joseph Niépce, in an 1827 note prepared for the Royal Society in London, styled his invention as 'spontaneous reproduction by the action of light'.[27] Daguerre in turn contended 'the DAGUERREOTYPE is not an instrument which serves to draw nature; but a chemical and physical process which gives her the power to reproduce herself'.[28] Talbot too plays on variants of this conceit: in the introduction to *The Pencil of Nature* (the title is already an example) he declared the images in his book were 'impressed by Nature's hand' (Fig. 3.1). The photograph he publishes of his ancestral home, Lacock Abbey (Fig. 3.2), appears to him as 'the first instance on record of a house painting its own portrait'. 'The plates of the present work are impressed by the agency of Light alone, without any aid whatever from the artist's pencil,' he notes. 'They are the sun-pictures themselves.'[29] The metaphoric idea of 'impression' and 'engraving by light', of a personified Nature wielding a pencil or the sun playing an artist, originates at the same time, and in direct relation to technologies devised and employed by mortals to make images with chemicals. These enunciations deny human authorship and install in its place the notion that something automatic, something indexical, something reflexive and unmediated happens whenever a photograph is taken.

Figure 3.1 William Henry Fox Talbot, title page, *The Pencil of Nature*

Figure 3.2 William Henry Fox Talbot, 'West Front of Lacock Abbey'

The confluence of photographic capacities associated with potent metaphors and analogies that were available to shape a collective understanding of those capacities resulted in the social formation of photography and the premise that photographs could be regarded as straightforwardly factual. Photography could function as a special kind of visual evidence – even in courts of law – because, for most of the medium's history, its testimony was believed, without significant qualification. Its reputation as a seemingly machine-authored or reflexive technology was sustained not simply because of how the process worked or what its images looked like, but because of a calculation we make that those who created and preserved photographs were not particularly motivated to distort the truth with them. Once that premise changes – once we know that Hollywood head shots are routinely airbrushed to remove imperfections, or begin to suspect that official Stalin-era portraits of commissars have been altered, or encounter a photograph of an unidentified flying object – photography's reflexive nature takes a back seat and our attention turns to the credibility of the photographer and the individuals or institutions presenting the evidence as implicitly factual.

The philosopher Siegfried Kracauer once noted that photography arrived in western culture at around the same time as the concept of historicism. Advocates of historicist thinking (he contends) believe they can explain any bygone event by tracing it back to its origins and then presenting its history as a steady unfolding of one thing after the next. 'That is, they believe at the very least that they can grasp historical reality by structuring the series of events in their temporal succession without any gaps.'[30] Kracauer finds in this an odd parallel with photography: if the photograph gives a continuum of all that resides before the lens at the moment of exposure – a spatial continuum – historicism aspires to represent past events as a temporal continuum, mirroring events as if some self-evident meaning would be revealed in their reproduction. Needless to say, he is sceptical of both. Human memory would make a better model for history writing than visuality, he argues, because, though full of gaps, it understands that events do not carry meaning within themselves. We remember only what is important, and the way we remember an event determines its meaning for us. Elsewhere he writes:

> A hundred reports from a factory do not add up to the reality of a factory, but remain for all eternity a hundred views of the factory. Reality is a construction. Certainly life must be observed if reality is to appear. Yet reality is by no means contained in the more or less random observational results of reportage; rather, it is to be found solely in the mosaic that is assembled from individual observations on the basis of comprehension of their import.[31]

Historians who use documents as sources – be they texts, maps, paintings or photographs – would be wise to embrace Kracauer's injunction, and not confuse factuality with truth.[1]

Notes

1 Joel Snyder has adeptly written about the status of the photograph as evidence in nineteenth-century legal theory, tracing its evolution from hearsay testimony to corroborating fact. See Joel Snyder, 'Res Ipsa Loquitur', in *Things that Talk: Object Lessons from Art and Science*, ed. Lorraine Daston (New York: Zone, 2004), 195–221.
2 The use of the terms 'photography' and 'the photograph' throughout this chapter betrays an unavoidable essentialism, which begs the question of how one defines what a photograph is. The problem may be illustrated with an example. The printed circuit board used in electronics are created using a light-sensitive resist to register the image (on film) of a pattern of lines that make up the circuit. The image is developed and the resulting resist is used to protect the metal layer of the board in an etching bath. The object this permanently registers an image (of the circuit) by means of light and light sensitive materials and chemistry. But is it a photograph? Does our definition of photography include printed circuit boards? Most would answer no, which only points out that what passes for an essential definition of the term is in practice only a normative one.
3 This is not true of digital photography, however, where the pixel is the smallest unit of management.
4 Charles S. Peirce, 'On a New List of Categories', *Proceedings of the American Academy of Arts*

and Sciences 7 (1868), 287–98, in *Peirce on Signs: Writing on Semiotic by Charles Sander Peirce*, ed. James Hooped (Chapel Hill, NC: University of North Carolina Press, 1991), 30.

5 Charles Sander Peirce, 'Logical Tracts, No. 2' (*c*.1903), in *The Collected Papers of Charles Sanders Peirce*, ed. Charles Hartshorne and Paul Weiss, vol. 4 (Cambridge, MA: Harvard University Press, 1931–58), 447.

6 As Snyder points out, a hammer thrown at a wall may make a hole in it, but the hole does not resemble the hammer. The notion of the index suggests a natural relation of referent to cause, but as Snyder observes, indices such as signet rings or cylinder seals will only make recognizable traces when used properly and as intended. Similarly, a boot dragged in the snow will produce a mark thereupon, and though an index, it will not be a bootprint nor necessarily provide a clue that it was a boot that was dragged and not some other object. See Joel Snyder, 'Picturing Vision', *Critical Inquiry* 6, no. 3 (1980): 507–8.

7 For a sense of just how fraught the concept of index remains, see 'The Art Seminar', *Photography Theory*, ed. James Elkins (New York: Routledge, 2007), 129–55.

8 Plato and Aristotle distinguished *mimesis* from *diegesis*, modes of showing from modes of telling. The framing intelligence implied by the photograph might be compared to a kind of narrator, specifically an invisible or omniscient narrator who tells through showing.

9 Roland Barthes, *Camera Lucida: Reflections on Photography*, trans. Richard Howard (New York: Hill & Wang, 1981), 5–6.

10 See Joel Snyder and Neil Allen Walsh, 'Photography, Vision, and Representation', *Critical Inquiry* 2, no. 1 (1975): 143–69.

11 Samuel Taylor Coleridge, *Biographia Literaria; or Biographical Sketches of My Literary Life and Opinions*, vol. 2 (London: Rest Fenner, 1817), 6.

12 The claim of *Camera Lucida* that every photograph is somehow co-natural with its subject is turned on its head by John Tagg, who argues that '*every* photograph is a result of specific and, in a sense, significant distortions which render its relation to any prior reality deeply problematic'. John Tagg, *The Burden of Representation: Essays on Photographies and Histories* (Amherst, MA: University of Massachusetts Press, 1988), 2, as cited by Sabine T. Kriebel, 'Theories of Photography: A Short History', in *Photography Theory*, ed. Elkins, 30.

13 By 'apparatus' here I mean both a technology and a related aesthetic developed to exploit the technology to certain ends; by 'maker' I mean both individual photographers and the cultural circumstances in which they are always embedded.

14 The art historian Ernst Gombrich calls this investment 'the beholder's share'. See Ernst Gombrich, *Art and Illusion: A Study in the Psychology of Visual Perception* (London: Phaidon, 1960).

15 'How to Produce Photographic Portraits without a Camera', *Anthony's Photographic Bulletin* 6, no. 17 (1875): 17.

16 Yi Gu, 'Dissecting the "Real": Photography and the Authority of Realistic Representation' in 'To Frame a View: Outdoor Sketching (*xiesheng*), Photography, and the Reinvention of Landscape Painting in China, 1912–1949' (PhD dissertation, Brown University, forthcoming). My sincere thanks to Yi Gu for her important analysis and for bringing the *Anthony*'s source to my notice.

17 Nadar [Gaspar Felix Tournachon], 'My Life as a Photographer' (1900), *October* 5 (1978): 6.

18 On this, see Geoffrey Batchen, 'The Naming of Photography: 'A Mass of Metaphor', *History of Photography* 17, no. 1 (1993): 22–32.

19   On this, see Lorraine Daston and Peter Galison, 'The Image of Objectivity', *Representations* no. 40 (1992): 81–128.

20   'Prospectus'; 'Introduction', *Daguerreotype* 1, no. 1 (1847): 5, as cited and discussed by Alan Trachtenberg, 'Photography: The Emergence of a Keyword', *Photography in Nineteenth-Century America*, ed. Martha A. Sandweiss (New York: Abrams, 1991), 17.

21   Ibid.

22   Edgar Allen Poe, 'The Daguerreotype', *Alexander's Weekly Messenger*, 15 January 1840, reproduced in *Classic Essays on Photography*, ed. Alan Trachtenberg (New Haven, CT: Leete's Island, 1980) 38. Emphasis in the original.

23   William Henry Fox Talbot, 'The Magic Mirror', *Legendary Tales in Verse and Prose* (London: 1830), reproduced in *Henry Fox Talbot: Selected Texts and Bibliography*, ed. Mike Weaver (Oxford: Clio, 1992), 37–9.

24   *Giphantie, A Tale by Charles-François Tiphaigne de la Roche* (Paris: Babylone, 1760) [quoted from 1761 London edition], 95–6.

25   Mayer and Pierson, *La Photographie consideree comme art et comme industrie* (Paris: Hachette, 1862), 8–11. See Mary Warner Marien, *Photography and its Critics: A Cultural History, 1839–1900* (New York: Cambridge University Press, 1997), 9–12.

26   See Steve Edwards, ' "Fairy Pictures" and "Fairy Fingers": The Photographic Imagination and the Subsumption of Skill', *The Making of English Photography: Allegories* (University Park, PA: University of Pennsylvania Press, 2006) for a compelling discussion of this context.

27   Joseph Nicéphore Niépce, 'Notice sur l'Heliography' 1827, MS. Gernsheim Collection, Harry Ranson Center, University of Texas, Austin, TX.

28   'An Announcement by Daguerre', *Image* 8 (March 1959): 34, as translated by Marien, *Photography*, 3.

29   William Henry Fox Talbot, 'Introductory Remarks', *The Pencil of Nature* (1844–6), facsimile reprint (New York: Da Capo Press, 1969), n.p.

30   Siegfried Kracauer, 'Photography', trans. Thomas Y. Levin, *Critical Inquiry* 19, no. 3 (1993): 421–36.

31   Siegfried Kracauer, *The Salaried Masses: Duty and Distraction in Weimar Germany* (1930), trans. Quintin Hoare (New York: Verso, 1998), 32.

4   Actuality and affect in documentary photography

by David Phillips

The category of documentary is one of the most long-standing in photography, as is the deployment of photographs as visual records or evidence. Yet what is meant by documentary is often unclear, and at times contradictory. In recent decades, particularly in the wake of semiotics and critical theory, many of the claims made for documentary photography have come to be seen as, at best, naive and confused or, at worst, as overtly duplicitous and, indeed, complicit with strategies of domination and disempowerment. Many of these critiques have validity, yet documentary endures, and an examination (rather than outright dismissal) of its various uses and connotations remains a productive exercise, not least because it provides access to the complexities of photography and, by extension, bears upon fundamental questions of how we both perceive and relate to 'reality'.

Index and truth

Since its beginnings, the photograph has been regarded as a direct imprint of reality, which presents an entirely faithful copy of the object or scene depicted. An assertion of the evidential value of the photograph (as visual record) underwrote the earliest conceptualizations of the medium and was made explicit in the first public announcement in 1839 of photography's discovery. Addressing the French lower house (the Chamber of Deputies), François Arago, Director of the Paris Observatory, posed the rhetorical question, 'Can this invention become practically useful?' before then answering himself by describing the daguerreotype's ability to produce 'faithful pictorial records ... [which] will excel the works of the most accomplished painters, in fidelity of detail and true reproduction of the local atmosphere'.[1] Although the term 'documentary' did not have wide currency in the 1830s, Arago was nonetheless promoting the role of the photograph as visual document by positing it as an image subordinate to its referent (as distinct from the image having its own intrinsic significance and value). For while acknowledging, if only in passing, the aesthetic properties of the daguerreotype, Arago primarily defined its value with reference to its 'practical use' and, related to this, to its utility as a tool for advancing other forms of knowledge (his own examples including archaeology and architectural restoration).

The daguerreotype was constrained by its being a unique (i.e. non-reproducible) image, but Arago's statement presciently foresaw photography's rapid conscription as 'evidence' across of a range of disciplines including anthropology, criminology and medicine. This deployment was itself predicated upon a growing differentiation, from the late eighteenth century onwards, between objective observation and subjective perception.[2] The opposing of science to art was fundamental to this division, and early assertions of photographic objectivity repeatedly rest upon the claim that a photograph was entirely free from the subjective distortions of human vision and draughtsmanship. Thus, describing the plates in *The Pencil of Nature* (1844–6), the British inventor of photography and Daguerre's rival, William Henry Fox Talbot, stated:

> They have been formed or depicted by optical and chemical means alone, and without the aid of anyone acquainted with the art of drawing. It is needless, therefore, to say that they differ in all respects, and as widely as possible, in their origin, from plates of the ordinary kind, which owe their existence to the united skill of the Artist and the Engraver.[3]

Such claims for the autogenesis of the photographic image abound in early accounts of the medium – commenting upon an image he had made in 1835 of his own house, Fox Talbot similarly noted that, 'this building I believe to be the first that was ever yet known *to have drawn its own picture*',[4] while Daguerre asserted that the daguerreotype was 'not merely an instrument which serves to draw nature … [but] on the contrary it is a chemical and physical process which gives her the power to reproduce herself'.[5] This promotion of photography as a mechanical and photochemical (and hence 'scientific') means of image-making that escaped the bias of human interests and perceptions continued well into the next century. In making his case for the intrinsic ontological realism of both photography and cinema, André Bazin could state that:

> Originality in photography … lies in the essentially objective character of photography … between the originating object and its reproduction there intervenes only the instrumentality of the nonliving agent. … All the arts are based upon the presence of man, only photography derives an advantage from his absence.[6]

Claims such as these for the objective, and hence also evidential, status of photographs are typically premised upon a belief in the medium's supposedly value-free neutrality – its (apparent) status, as described by Roland Barthes, as 'a message without a code'.[7] Above all, it is the indexical connection between the photograph and its referent,[8] whereby the image is causally dependent upon the object it represents (e.g. a fingerprint), that underwrites claims made for photography's innate mimetic realism. At its most extreme, this line of argument asserts that a photograph is equivalent to, or a substitute for, the referent it depicts with the result that any distinction between the two is effaced – as Bazin puts it, 'The photographic image is the object itself … [the image] shares, by virtue of the very process of its becoming, the being of the model of which it is the reproduction; it *is* the model.'[9]

Although the indexical connection between image and referent is unique to (non-digital) photography, in that a photograph is produced by exposing a light-sensitive surface to light, photography is a cultural practice and not a natural process. As such, it needs to be understood as an interpretive rather than as a transcriptive medium. Not only is there little correspondence between the constantly shifting focus and continuity of binocular human vision and the frozen photographic image, but camera technology has specifically been evolved also to replicate the pictorial conventions of perspectival space (including depth of field, sharp focus, and framing), thereby reinforcing a belief in the photograph as a transparent window onto the world. The actual process of taking a photograph is also, to some degree, always a highly orchestrated event. A photographer (often in conjunction with a sitter) makes a series of decisions relating to lighting, exposure time, the choice of film type and the placement of the camera, and the negative then has to be developed and the print (itself determined in part by the choice of printing paper) further manipulated by enlarging or cropping. But although the claim for photographic objectivity has come to look increasingly naive, the recurrent identification of the photograph with visual truth has been fundamental to photography's ascendancy and power. Not only have particular values such as clarity, sharpness and transparency become associated with photography and the 'photographic', but also this way of thinking about photography remains a feature of our everyday language (as typified by figures of speech such as 'the camera never lies' and 'photographic realism'). These associations in turn have reinforced the implication that specifically pictorial values (such as accuracy) have moral connotations (such as honesty and integrity) – a conflation of pictorial properties and ethical attributes that has been especially important for documentary photography.

But rather than posit objectivity as an innate attribute of mechanical image-making techniques, we might ask, instead, both why and how it is that these mechanical processes have come to be seen as objective. As Pierre Bourdieu observed, 'Photography is considered to be a perfectly objective and realistic recording of the visible world because (from its origin) it has been assigned *social uses* that are held to be "realistic" and "objective".'[10] This view is echoed by Joel Snyder in his claim that:

> Documentary is a classificatory category that is established by use and not by essential character. Thus, calling photographs documentary is not contingent upon the assignment of some special quality of accuracy to them; it is providing them with certain uses.[11]

In short, as distinct from its being intrinsically 'realistic', photography is *required* to be seen as objective in order for it to perform particular representational tasks. But a presumption of indexicality alone is no guarantee of veracity or, indeed, of meaning. As Walter Benjamin, citing Bertolt Brecht, observed in the early 1930s (the heyday of documentary's codification and authority), 'less than ever does a simple *reproduction of reality* express something about reality'.[12]

More recently, John Tagg has noted: 'Ask yourself, under what conditions would a photograph of the Loch Ness Monster or an Unidentified Flying Object become acceptable as proof of their existence?'[13] For Tagg, and other commentators, the veracity of a photograph is secured extrinsically. Most immediately, this entails the framing of a photograph by some kind of textual support (titles, captions, editorial commentary, etc.) that serves to limit the potential openness (polysemy) of the photograph. As described by Roland Barthes:

all images are polysemous; they imply, underlying their signifiers, a 'floating chain' of signifieds, the reader is able to choose some and ignore others. Polysemy always poses a question of meaning and this question always comes through as dysfunction. ... Hence in every society various techniques are developed to fix the floating chain of signifieds in such a way as to counter the terror of uncertain signs; the linguistic message is one of these techniques.[14]

Taking up Barthes's analysis (including his deliberate reference to the rhetorical functioning of images), Allan Sekula has described how meaning is fixed discursively:

A discourse can be defined as an arena of information exchange, that is, as a system of relations between parties engaged in communicative activity. In a very important sense the notion of discourse is a notion of limits. That is, the overall discourse relation could be regarded as a limiting function that determines the very possibility of meaning. ... A discourse, then, can be defined in rather formal terms as the set of relations governing the rhetoric of related utterances. The discourse is, in the most general sense, the context of the utterance, the conditions that constrain and support its meaning, that determines its semantic target.[15]

This regulative assemblage of texts and discourses is frequently bound up with institutions such as publishers, the police, government agencies, and welfare administration, and it is these institutional forces, not the photograph's intrinsic realism, that invest it with the authority of the real. Bourdieu has also pointed to photography's broader tautological function here in that typically its serves to reiterate and confirm already existing (and dominant) representations:

only in the name of a naïve realism can one see as realistic a representation of the real which owes its objective appearance not to its agreement with the very reality of things ... but rather to conformity with rules which define its syntax within its social use, to the social definition of the objective vision of the world; in conferring upon photography a guarantee of realism, society is merely confirming itself in the tautological certainty that an image of the real which is true to its representation of objectivity is really objective.[16]

Photographs may present themselves as value-free and disinterested, but we should regard them instead as constructions that naturalize particular versions of reality. Moreover, as Sekula comments:

All communication is, to a greater or lesser extent, tendentious; all messages are manifestations of interest. ... The overwhelming majority of messages sent into the 'public domain' in advanced industrial society are spoken with the voice of anonymous authority and preclude the possibility of anything but affirmation.[17]

The supposed transparency and apparent literalness of photography underwrite its 'anonymous authority' making it highly amenable to this task of affirmation, particularly when photographs are presented as evidence. Not surprisingly, photography was rapidly adopted by a range of governmental agencies and emergent social sciences keen to harness its 'realism'. As John Tagg, deploying a Foucauldian analysis of photography, has argued, 'What gave photography its power to evoke a truth was not only the privilege attached to mechanical means in industrial societies, but also its mobilization within the emerging apparatuses of a new and more penetrating form of the state.'[18] Connotations of truthfulness were secured, in part, for the evidential and scientific use of photographs by means of a repertoire of visual techniques – including frontal or profile poses in conjunction with neutral backgrounds and standardized pictorial formats (to aid visual comparison) – that were frequently supplemented by accompanying statistical data and measurements. Signifying 'objectivity', such pictorial conventions served also to legitimize particular power relations as the subjects of the photographs (typically, 'colonized' races and the socially marginalized who are presented as specimens or types) became the objects of a disciplinary gaze. As Abigail Solomon-Godeau has observed:

> We must ask ... whether the documentary act does not involve a double act of subjugation: first, in the social world that has produced its victims; and second, in the regime of the image produced within and for the same system that engenders the conditions it then re-presents.[19]

The visual encoding of photographic neutrality (for example, the evolution of the police 'mug shot' and, later, of the passport photograph) helped secure the evidential status of the photograph, but this was a process prompted also by debates, even controversy, surrounding the use of photographs as evidence. An early instance of this arose from the use of photographs by the Barnardo homes in London. In 1871 Dr Thomas John Barnardo had opened the first of several homes in the East End of London dedicated to rehabilitating destitute street children and, in order to advertise his success and generate funds, he had established a photographic studio to produce 'before' and 'after' *cartes-de-visite* portraits of his young wards. However, in 1876 Barnardo was accused by the Rev. George Reynolds of various counts of misconduct. In particular, Reynolds claimed that one photograph, *The Raw Material As We Find It* (Fig. 4.1) was an 'artistic fiction' and that a number of other photographs were also wilfully 'deceptive'. After taking the case to the Arbitration Court the following year, Barnardo was cleared of criminal intent (although the Court did uphold the charge of 'artistic fiction') and he subsequently published a defence of his photographs. Barnardo conceded that the title, *The Raw Material As We Find It*, could possibly 'lead the reader to suppose that the boys depicted were literally found *together* in a group, and not separately, as was actually the case', but added that, 'their individual condition on the night when they were rescued ... was proved to have been precisely the same as is shown and described in the photograph.'[20] In response to the charge that several other photographs were deceptive, because the children's clothes had been deliberately torn and props had been used, Barnardo further asserted that the photographs were always intended to be 'representative or typical' portraits. The question of whether photographed subjects were to be depicted as specific individuals or as generic (and frequently anonymous) 'types' would be a recurrent one, particularly in social documentary photography. But the Barnardo case is an example, also, of how the authenticity of documentary photography was open to being challenged, just as the production of the photographs themselves highlights the question posed by Solomon-Godeau, as the images were intended for viewers who had little, if any, direct connection with the photographed subjects themselves. In effect, it was the values of that audience that determined the photographs' currency and meaning.

Figure 4.1 Thomas Barnes and Roderick Johnstone, *The Raw Material As We Find It, 1875*

Document, documentary and art

Despite the reiteration (in various guises) of the essentialist claim for the innate documentary or transcriptive identity of photography, it remains nonetheless only a claim. Ultimately, the broad association of photography and documentary is of little value for understanding individual photographs or for assessing the specific attributes of documentary. A photograph can undoubtedly acquire documentary or evidential value, but it may not have been made with that purpose in mind – for example, a nineteenth-century photograph of a city street may provide valuable information about urban hygiene and traffic density, but this information may have been incidental to the original purpose in producing the image (for example, the promotion of the city as a historic location or as a tourist venue). Similarly, a portrait may be a useful documentary record of contemporary fashions, yet it is unlikely that this aspect of the photograph was originally the primary consideration. Moreover, not only does an assertion of the inherently documentary nature of photography efface the differences between various types of photographs (including their contexts and functions), with the result that all photographs effectively become documentary, but also it overlooks the variations within the documentary tradition itself.

Any reference to a 'documentary tradition' necessarily raises the question of genre classification, which continues to remain an issue in photography, particularly for galleries and museums seeking to secure an artistic pedigree for particular photographers. Influential curators and photo-historians – in particular, those associated with the Museum of Modern Art (including Edward Steichen, Beaumont Newhall and John Szarkowski) – have sought to distinguish between various genres in photography. Their categories have included art photography, documentary, photojournalism, but many others can be added (fashion, advertising, portraiture), as well as distinctions between professional and amateur photography, each of which has its own genres (e.g. family albums and tourist snaps that comprise most 'amateur' photography). These hierarchical divisions may have shaped both the production and consumption of photographs (with a privileging of art photography), but they are by no means clear-cut, and photographs can frequently move between supposedly distinct categories. The critical assessment of the French photographer, Eugène Atget, is a case in point, as his elevation, in recent decades, from jobbing artisan to creative artist demonstrates how photographs can acquire new values (including both aesthetic and monetary).[21] This shift in perception was in part achieved by detaching Atget's photographs from their original context and function. But by returning Atget to his specific historical situation, it is possible to track significant developments in the photographic document, in particular the evolution in the 1920s and 1930s from 'document' to 'documentary' – a shift from a functional understanding of the document (defined by its practical utility) to an aesthetic understanding of documentary (as a genre defined as much by style and expressive effect).

Atget took up photography in the late 1880s working with glass plates and a large bellows camera mounted on a tripod, despite the availability of hand-held cameras, and within a few years was selling his prints to an eclectic range of clients that included artists, decorators, designers, architects, shopkeepers, publishers, government archives, libraries and museums. As a commercial photographer, Atget's oeuvre reflected his clients' needs, and he accumulated an extensive archive of architectural photographs that included street views, houses, courtyards, churches, statuary, staircases, doorways, interiors and decorative features (Fig. 4.2). In addition to this stock of prints, Atget systematically photographed specific subjects (including street traders, markets, churches, bars, shop fronts and window displays) and particular localities in and around Paris, including the 'zone' on the city's outskirts and the pre-Revolutionary royal castles and gardens of Versailles, Saint-Cloud and Sceaux.

Figure 4.2 Eugène Atget, *Au Tambour, 63 quai de la Tournelle, 1908*

Atget referred to himself as a 'documenter' and advertised his photographic service as providing 'documents for artists'. But what he and his contemporaries meant by 'document' was extremely vague, as a document was not defined by a particular genre, medium, format or style, but was instead defined by its having a practical function or use. As Molly Nesbit notes:

> a document was a study sheet. Its beauty *was* secondary; use *did* come first ... an architectural photograph would be called a document, as would a chronophotograph, a police i.d., or an X-ray. They all had one thing in common: all of them were pictures that went to work.[22]

More specifically, the document was a means to an end or a stage in a production process as in, for example, a technical drawing or blueprint. In order to most effectively discharge this function, a document had to meet specific conventions and expectations. Not surprisingly, clarity and legibility were especially valued, although not as aesthetic ends in themselves, but so as to facilitate the specialist's comprehension of technical information.

The particular suitability of photography for this kind of task had been recognized early on – as typified by Arago's statement in 1839 – and Atget's understanding of the photographic document maintained this instrumentalist understanding of photography as mechanically produced visual aid. Yet, although the document was essentially functional, it was also evolving and acquiring status. As Nesbit further observes:

> in the late nineteenth century the document was fast becoming something more than a preparatory step for buildings, paintings, and ornamental details. Whether drawn or photographed, the document was playing an increasingly important role in the elaboration of scientific and historical proof. It became a standard way of expressing knowledge; it became a means to knowledge; and it put together pictorial forms of knowledge, though they were not yet taken up as aesthetic forms and exploited for their own sake. That would come later.[23]

Atget may have regarded his photographs as neutral pictorial documents, judged principally in terms of the visual information (content) they conveyed, but he was also an important transitional figure in the history of documentary photography. In particular, the critical reception of his work contributed to a shift from a strictly functional (non-aesthetic) understanding of the document to an aesthetic understanding of documentary. Atget's photography was predicated upon practical utility (as articulated by Arago). Yet, by rejecting both the conventions of picturesque and the soft-focus pictorialism of late-nineteenth-century art photography, Atget also presaged the 'straight' photographic aesthetic, and engagement with popular culture, that were integral to interwar documentary. Claimed as a precursor by the Surrealists, who hailed him as 'naive' primitive, Atget was also greatly admired by American documentary photographers, including Berenice Abbott (who acquired his estate and eventually presented it to the Museum of Modern Art) and Walker Evans, who commended what he described as Atget's 'lyrical understanding of the street [and] trained observation of it'.[24] Although working in a documentary mode, Evans' own practice entailed scepticism about documentary, and he referred instead to 'documentary style', making an emphatic distinction between a document and art:

> The term should be *documentary style*. An example of a literal document would be a police photograph of a murder scene ... a document has use, whereas art is really useless, therefore art is never a document, though it can certainly adopt that style.[25]

Evans' reference to 'documentary style' was not necessarily shared by his contemporaries, who espoused more conventional notions of realism, but he was no more than acknowledging that many of the features coded as 'documentary' by the 1930s (clarity, frontality, directness, etc.) were themselves formal and stylistic devices and that the apparent absence of style in documentary had itself now become a signifier connoting objectivity and neutrality.

Fact and feeling: the human document

If the recording and presentation of visual information and evidence are necessary to documentary photography, these tasks do not in themselves provide a sufficient definition of the term 'documentary', nor do they exhaust what is at stake in the genre. Atget's functionalist method and neutral style represented one possibility, but the aim of much documentary photography is to effect a change in perception and thereby persuade its audiences of the need for social action (typically the amelioration of a particular social situation). Not surprisingly, the line separating information and instruction from partisan polemic is not always clear – as the British documentary filmmaker, John Grierson, acknowledged, 'there is hardly any avoiding the accusation of propaganda'.[26] To varying degrees, documentary photography is didactic, and this function of education and persuasion is implicit in the evolution of the words 'document' and 'documentary'. First used in the early nineteenth century by the British philosopher, Jeremy Bentham, in his 'Rationale of Judicial Evidence' (1827),[27] the adjective 'documentary' was derived from the noun 'document', itself traceable to several centuries earlier, which specifically referred to a written record. However, these connotations of evidence, proof and record (which clearly relate to common beliefs about photography) are relatively modern. The Latin word *documentum* principally referred to a lesson, and although the earlier meanings of 'document' (lesson, instruction and even warning) have largely been superseded by modern legalistic connotations, older usages remain implicit in 'documentary'.

The combination of evidence and instruction that characterizes much documentary practice necessarily entails various rhetorical and aesthetic techniques that combine fact with feeling, information with affect, and factuality with polemic. Arguably, these combinations (for instance, the claim to neutrality while simultaneously using photography as a vehicle for social advocacy) are unsustainable, yet they remain essential to documentary practice. For example, despite his preference for dramatic symbolic expression over naturalism in films such as *Drifters* (1929), John Grierson stated that, 'Documentary was from the beginning ... an anti-aesthetic movement.'[28] Yet Grierson's own films embody a method that he himself defined as 'the creative treatment of actuality'. Although this phrase awkwardly combines terms that do not readily coexist and, indeed, potentially contradict each other – 'creative' (artistic) treatment or licence and 'actuality' (reality as it is) – this juxtaposition exemplifies a dual strand within documentary. Grierson's recognition of an expressive dimension to documentary has been echoed by William Stott's observation, in his landmark study of 1930s documentary media, that while documentary is grounded within 'the genre of actuality', its purpose is to 'educate one's feelings'. Stott develops his assertion further, noting that: 'We understand a historical document intellectually, but we understand a human document emotionally. In the second kind of document, as in documentary and the thirties' documentary as a whole, feeling comes first.'[29] As Roy Stryker, editor of the Farm Security Administration Survey (discussed below) commented, 'Truth is the objective of the documentary attitude,' adding that, 'A good documentary should tell not only what a place or a thing or a person *looks* like, but it must also tell the audience what it would *feel* like to be an actual witness to the scene.'[30]

Stryker's understanding of documentary affect turns, in part, upon the viewer's emotional identification with a specific situation or with particular individuals or groups. But his reference to the viewer's role as witness (rather than, for example, as imaginary participant) introduces another dimension – namely, the presentation of actuality as spectacle. As Elizabeth Cowie has argued:

> In recording actuality, photography and cinematography address two distinct and apparently contradictory desires. On the one hand there is a desire for reality held and reviewable for analysis as a world of materiality available to scientific and rational knowledge, a world of evidence confirmed through observation and logical interpretation. ... On the other hand there is a desire for the real not as knowledge but as image, as spectacle.[31]

The photographic document may present itself as purged of subjectivity, yet its very existence might be regarded not only as a manifestation of 'a basic psychological need' (Bazin) for the real, but also as being invested with decidedly 'human' emotions.

Case study: social documentary in America

Stott's stress upon 'the primacy of feeling' foregrounds the instrumentalist stance of social documentary in particular. Lewis Hine, arguably the quintessential social documentary photographer and who, like Grierson, had a background in social research, summarized this stance when he observed that his photographs were intended 'to show the things that had to be corrected' and to provide 'a lever ... for the social uplift'.[32] For Hine, as for other practitioners of 'social photography', the photograph could provide compelling visual evidence that would support calls for social reform, and I wish to turn now to examine some examples of American reform photography with reference to the issues discussed above. Examining just a small selection of photographs – all produced in just one country within only a few decades – reveals a range of documentary practices which, despite their divergent premises and methods, all made use of photographs as evidence. But there are commonalities, too, not least in the appeal to the viewer's emotional response, whether it be voyeuristic fascination, moral outrage or humanist empathy.

One of the earliest, and most influential examples, of documentary photography in America was *How the Other Half Lives* (1890), a book which not only was a landmark in the use of photography for social reform, but also prefigured many of the techniques of investigative journalism. An immediate success (with eleven printings in just five years), *How the Other Half Lives* was written by a Danish-born immigrant, Jacob A. Riis, who arrived in New York in 1870, eventually finding permanent employment as a police reporter for the *New York Tribune* and, subsequently, for the *Evening Star*. At that time, the headquarters of the New York police was on Mulberry Street in the midst of the notorious Lower East Side slum district, and Riis's knowledge of the area and its inhabitants formed the basis of his journalistic campaigns exposing the degradations of life in the slum tenements.

From 1877 to 1887, Riis lectured on New York's slums, but it was the invention of flash photography in 1887 that provided him with a powerful new resource. In addition to the street maps and building plans he already used, Riis was now able to show images of previously hidden tenement interiors (Fig. 4.3) and 'magic lantern' slideshows became an integral component of his lectures. Riis himself photographed only from 1888 to 1889, and other (unacknowledged) photographers, including Richard Hoe Lawrence and Dr Henry G. Piffard, took many of the images credited to Riis himself – a situation that necessarily raises questions over authorship. The lectures themselves were dramatic performances, given to largely middle-class audiences, and Riis's use of photography was motivated as much by entertainment and spectacle as it was by humanitarian reform. At times, the lectures included Bible readings and prayers, and music was also used for dramatic and emotional effect. Riis's commentary on his photographs combined moralizing polemic with melodramatic effect as the photographs were simultaneously presented as both objective evidence and as visual entertainment. Riis's commentary, and the text of *How the Other Half Lives*, also included his anecdotes about incidents that occurred while visiting the slums and about specific individuals and directly shaped the specific meanings of the photographs by providing an authenticating and interpretive framework for them that guided his audience towards his own values and beliefs. As described by Maren Stange:

> The lectures embedded the evidentiary image in an elaborate discourse offering simultaneous entertainment and ideology, and from this the photograph, no matter how seemingly straightforward, never stood apart. ... Riis made each image a rich carrier of specific ideological messages.[33]

This heterogeneity of genres in Riis's work would not necessarily have presented a difficulty for his audience, who had a broad and flexible understanding of realism and who could, for example, readily accept the authenticity of a deliberately staged photograph – as many of Riis's photographs were. Moreover, the combination of evidence and entertainment had its precedents: Daguerre himself may have described his own invention as 'an instrument which serves to draw nature', but he had previously been a successful set designer, and then owner of the Diorama, and was to promote the daguerreotype as a novel curiosity which the 'leisured class will find ... a most attractive

Figure 4.3 Jacob A. Riis, *Five Cents a Spot, c.1889*

occupation'.[34] But Riis's lectures also played to his audience's social anxieties and to their familiarity with existing visual representations of urban crime and poverty. In an interview in 1888, Riis observed that: 'The beauty of looking into these places without actually being present there is that the excursionist is spared the vulgar sounds and odious scents and repulsive exhibitions attendant upon such a personal examination.'[35] Not only did *How the Other Half Lives* offer its middle-class readers the vicarious thrill of seeing parts of New York that they would never normally encounter (indeed, would deliberately avoid), but also the book both echoed and reinforced existing perceptions of the poor, as a number the photographs deliberately replicated the often-sensationalist drawings of vice and debauchery that illustrated contemporary publications such as *The Daily Graphic* - an instance of photography 'objectively' affirming representations already in circulation.

Despite, but perhaps because of, his own experience of poverty as a newly arrived immigrant, Riis strongly identified with self-reliance and enterprise, and repeatedly stressed the virtues of family and work.[36] Although using the camera to bring his audience up close to the urban poor, Riis's lectures and books reinforced the social divide between his 'excursionist' middle-class readership and the inhabitants of the slums who, framed within discourses derived from colonialism and tourism, are positioned on the other side of a social gulf. Not only does Riis rely upon national and ethnic stereotyping, but also he shows little concern for the individual lives of his sitters, as individual and ethnic differences become subsumed within a cumulative depiction of a generic and animalistic slum dweller existing beyond the reach of respectable conventions and morality.

The combination of the sociological and the sentimental, and of urban surveillance and tourist spectacle in Riis's photographs makes it difficult to present him as an obvious precursor of later documentary (just as, for example, his work does not conform to subsequent conventions of 'straight' photography), while it can also make it difficult to pin down his attitude to the poor, which can appear to be both harsh and sympathetic. From one perspective, his images would seem to endorse a view of the poor as dissolute and feckless, yet the images can also be read as revealing the debilitating effects of poverty. Often judgemental, and at times racist, Riis did nonetheless stress the causal connection between human behaviour and living conditions and, as such, *How the Other Half Lives* contains elements of nineteenth-century paternalism while prefiguring more modern discourses of environmental health. This connection between poverty and behaviour is in part established by the pictorial style of the photographs themselves. Frequently dark, claustrophobic and full of random clutter (including sprawled and slumped bodies), Riis's images function as visual analogues of the chaotic disorder of the slum interiors they depict, while their lack of technical sophistication (poor focus, overexposure and flare, the lack of a compositional centre and the graininess of the prints themselves) creates an effect of authenticity.

Riis's frequently intrusive and voyeuristic representation of slum life was challenged by his effective successor, Lewis Hine, who condemned the sensationalism of what he referred to as 'yellow photography', noting that, 'while photographs may not lie, liars may photograph'.[37] Not only was Hine the most significant reform photographer of the Progressive Era, but also he is arguably the quintessential social documentary photographer. As such, his work reveals both the achievements and the limitations, even contradictions, of social documentary photography. In particular, Hine's idealization of the poor was often predicated upon their passivity, individually and collectively, both before the camera and as social agents, while the socially critical aspects of his earlier activist work was increasingly compromised by his need, as a working photographer, to find commissions, including those from industrial corporations. Despite Hine's own motivations for taking up the camera, he often had little control over the final use, and thus meaning, of his photographs.

Trained as a sociologist and teacher at the University of Chicago, Hine subsequently taught for several years at the Ethical Culture School in New York. Here he began to use photography as a teaching aid, and this pedagogic purpose is evident in his photographs of European immigrants disembarking at Ellis Island. Migrants had also been photographed by Riis, but whereas Riis had played upon the social gulf that separated them from the viewers of his photographs, Hine sought instead to counter the widespread suspicion of newly arrived immigrants by emphasizing a common humanity between them and the viewer. Hine achieved this by means of various formal techniques for, while espousing a belief in the photograph's innate 'realism if its own', he was also alert to photography's aesthetic mediation of what it depicts. Stating that a photograph 'is often more effective than the reality', Hine developed a repertoire of pictorial rhetorics that were to typify 'straight' documentary. Rejecting the melodramatic lighting and visual clutter that characterized Riis's photography, Hine instead cultivated a style based upon evenly distributed (and preferable natural) light, sharp definition, clarity of detail, and a centralized frontal composition which, by bringing his subjects close to the camera, allowed them to fill the frame. Above all, it is the direct mutual engagement between Hine and his subjects that in turn elicits the viewer's empathetic response. In contrast to the averted gazes, stupefied poses and claustrophobic interiors in Riis's photographs (the signifiers of unbridgeable social and moral difference), the physical proximity, straight-forward frontal poses and reciprocal eye contact in Hine's pictures seem to offer the viewer the possibility of a free exchange of gazes – and, by extension, a direct personal contact – that transcends class and ethnic difference.

Writing in 1938, Beaumont Newhall was alert to the connection in Hine's work between aesthetics, emotion and documentary:

> These [Hine's] photographs were taken primarily as records. They are direct and simple. The presence in them of an emotional quality raises them to works of art. 'Photo Interpretations', Hine calls them. He might as well call them 'documentary photographs'.[38]

In a later account of the photographs, Newhall reiterated this reading, highlighting Hine's combination of objective and subjective elements:

> Hine realized ... that his photographs were subjective and, for that very reason, were powerful and readily grasped criticisms of the impact of an economic system upon the lives of underprivileged and exploited classes. He described his work as 'photo-interpretations'. The photographs were published as 'human documents'. Unbothered by unnecessary details ... throughout his pictures the harmony can be felt.[39]

But in addition to their formal and stylistic qualities, and in keeping with his observation that photography belonged to the tradition of 'the ancient picture writers' and 'continues to tell a story packed into the most condensed and vital form', many of Hine's documentary images also suggest symbolic and narrative possibilities. In *Young Russian Jewess at Ellis Island* (Fig. 4.4), the situation of the young woman intentionally recalls the original Pilgrim settlers, as their flight from religious persecution is recreated in a contemporary setting. So too, despite her Jewish identity, the circle of light behind the woman evokes Christian imagery of the halo and the Madonna, which are further reinforced by the picture's compositional resemblance to a Russian icon. Symbolic resonances such as these complement the echoes of familiar narratives, including the Wandering Jew or the young Virgin Mary seeking shelter, which, although not overt, further humanize the young woman as she moves from the Old World into the New and from the past into the future.

Hine's self-styled 'uncomplicated' visual style had an overtly utilitarian and evidential premise – as he stated, 'I wanted to show the things that had to be corrected. I wanted to show things that had to be appreciated' – and is particularly apparent in the hundreds of photographs he produced as a staff photographer for the National Child Labor Committee (NCLC). Often posing as an inspector of machinery (rather than of working conditions), Hine photographed children working in cotton mills, coal mines,

Figure 4.4 Lewis Hine, Young Russian Jewess at Ellis Island, 1905

canneries and in the newspaper industry, and his images were used extensively in NCLC publications, exhibitions and slideshows as part of its campaign for the regulation of child labour. However, Hine's undoubted achievement in giving a human face to anonymous immigrants or child workers was, in large part, predicated upon an optimistic and rationalist belief that, once social injustices were made visible (in effect, through the indexical and moral force of the photograph), viewers would feel compelled to act and that reform would necessarily follow. In this, Hine perhaps shared the stance of the Progressive reform movement which, while focusing on the worst excesses of exploitation, offered little in the way of systemic critique. Indeed, as George Dimock has suggested, 'Our experience of these images may include, perhaps, the fantasy that we can "save" (or have "saved") those children.'[40]

By the 1930s the language of moralizing paternalism that had framed documentary photography, such as *How the Other Half Lives*, had been displaced (to an extent) by welfare initiatives and interpretive models shaped by the social sciences, especially sociology and economics. This mutually supportive connection between documentary and the procedures of the social sciences had characterized Lewis Hine's work, particularly his work as staff photographer on the six-volume *Pittsburgh Survey* (1909–14), although much of the authority of his photographs also rested upon Hine's own status as personal witness and moral voice. But Hine also worked for corporate interests, and the harnessing of photography to governmental and bureaucratic agendas was to be a defining feature of documentary in the 1930s and 1940s. During these decades, photography not only became the most significant facet of the mass media, but also was increasingly professionalized, as photographers, offering their services for hire, were obliged to supply images that met their clients' requirements. This conscription of documentary 'realism' in the service of governmental agendas, and the concomitant recognition of the mass media as an instrument of state policy, indeed propaganda, were especially evident in the work of the Farm Security Administration (FSA) that operated from 1937. One of the many government agencies established by President Roosevelt as part of his New Deal initiative to counter the Great Depression, the FSA (originally titled the Resettlement Administration, founded in 1935, but renamed two years later) provided subsidized loans to farmers, supplied equipment and farm supplies, assisted with relocation, and promoted land conservation. The Director of the RA/FSA, Rexford G. Tugwell, had previously co-authored a book, *American Economic Life and the Means of its Improvement* (1925), that had been extensively illustrated by photographs (many by Hine), and was alert to the potential of photography for publicizing the plight of rural communities and for promoting federal government initiatives. With this purpose in mind, he established a Historical Unit, under the direction of Roy E. Stryker, to produce and supply pictures for newspapers and magazines, such as *Survey Graphic*, *Life* and *Look*, and for touring exhibitions.

Based in Washington DC, Stryker appointed a number of photographers – initially, Arthur Rothstein, Carl Mydans, Ben Shahn, Walker Evans, Dorothea Lange, Russell Lee followed by Marion Post Wolcott, Jack Delano, John Vachon, John Collier and Gordon Parks. Stryker provided them with detailed 'shooting scripts', supplemented by correspondence, outlining not only what was to be photographed but, often, how designated subjects were to be photographed. Formerly an economics professor at Columbia, Stryker himself was not a photographer but as co-ordinator, editor and final arbiter (who famously destroyed negatives by punching a hole through them) of the vast FSA archive, he can be credited as its author. Yet although the immediate aim of the Historical Unit was to supply photographs that could be used by Washington to mobilize support for government relief programmes (and, from 1941, for the war effort), the survey in fact became the most ambitious government-funded documentary record of a nation which, upon its completion, comprised approximately 165,000 negatives.

The photographs produced by the FSA were presented as objective records, yet they were also intended to publicize and mobilize support for New Deal policies. Commenting on a photo-essay published in 1939 in *U.S. Camera*, Edward Steichen emphatically identified documentary with direct emotional and moral appeal:

> Have a look into the faces of the men and the women in these pages. Listen to the story they tell and they will leave with you a feeling of a living experience you won't forget; and the babies here, and the children; weird, hungry, dirty, lovable, heart-breaking images; and then there are the fierce stories of strong, gaunt men and women in time of flood and drought. If you are the kind of rugged individualist who likes to say 'Am I my brother's keeper?' don't look at these pictures – they may change your mind.[41]

This coexistence of objectivity and persuasion echoed deeper ideological imperatives, even contradictions, within the FSA. For while documenting the breakdown of rural life and eliciting sympathy for a dispossessed rural population (especially the poor tenant farmers and sharecroppers most affected by falling prices and land degradation), the FSA was also an agent of the rural modernization (for example, the mechanization of farming) that in part contributed to the plight of farmers. For despite its ostensible commitment to the farming families, the FSA did not in fact support independent and freehold farmers, but instead endorsed the large-scale farming and centralized planning that had partly contributed to the forced migration of rural families. This tension within FSA policy was resolved in part by the pictorial strategies adopted by the Historical Unit as attention to the actual causes of economic depression and social dislocation was displaced by a humanitarian response based upon shared universal values. Typically – and as described by Steichen – the individuals depicted in FSA photographs are presented as generic types and emblematic symbols representing transcendent qualities such as stoic endurance (rather than anger) or as helpless victims whose situation is more a result of fate or natural disaster than a consequence of economic policy. Images of individuals and families largely resigned to their fate vastly predominate over images of collective action. While much FSA photography invites identification between viewer and subject, this is achieved by picturing subjects who are largely excised of individuality and historical specificity.

Dorothea Lange's photography exemplified the liberal humanism of much FSA photography, but her work also shows a particular attention to the psychology of the sitter as revealed through the expressive body. This focus upon gesture and pose combines actuality with expressive affect suggesting a metaphorical quality and symbolic resonance beyond literal reportage. As Roy Stryker observed of Lange's *Migrant Mother* (Fig. 4.5), an image which he regarded as the definitive FSA photograph, 'You can see anything you want to in her',[42] a comment echoed by Lange's own observation that photography should speak 'in terms of everyone's experience'. *Migrant Mother* has

Figure 4.5 Dorothea Lange, *Migrant Mother, 1936*

become of one of the iconic images of 1930s documentary, yet the sequence of photographs culminating in the final image reveal a deliberate management of meaning on Lange's part. By moving in close to the group, distracting detail (including the family's scattered belongings) and the particularities of the location were removed, thereby allowing the image to resonate on a more abstracted and generic level, while greater prominence was given to the (unnamed) mother's expression and the young children's poses. The family in fact included seven children (not all of whom were present) and no father is present but, importantly, Lange removed the oldest daughter from the final image so as not to offend middle-class viewers of the photograph and thereby compromise its purpose of justifying government aid to such families. As James Curtis explains, 'Lange was undoubtedly influenced by prevailing cultural biases. A family of seven children exceeded contemporary social norms.'[43]

Despite the rapid iconic status acquired by *Migrant Mother*, Lange did not include the photograph in *An American Exodus: A Record of Human Erosion* (1939), the book she published with her economist husband, Paul Taylor. Combining photographs and text, the book documented the mass displacement of the population from drought-stricken Oklahoma. *American Exodus* was in part a response to another book, *You Have Seen their Faces* (1937), a joint collaboration between the photographer, Margaret Bourke-White, and the writer, Erskine Caldwell. *You Have Seen their Faces* had been commercially successful, but had also met with criticism on the grounds that it both caricatured and sentimentalized the sharecroppers and that Caldwell had taken creative liberties with the statements made by the farmers themselves. In *American Exodus*, Lange and Taylor, despite their empathy for the plight of migrant families, sought to avoid the perceived excesses of *You Have Seen their Faces* cultivating instead a tone of sober objectivity by compiling facts and statistics alongside the pictures, while Lange deliberately avoided using her most expressive images (including *Migrant Mother*). Seeking to balance objective record with emotive appeal, evidence with opinion, and factuality with aesthetic effect, the book's production entailed a finely calibrated mix of the potentially contradictory imperatives that continue to shape documentary practice.

Notes

1 Dominique François Arago, 'Report', in *Classic Essays on Photography*, ed. Alan Trachtenberg (New Haven, CT: Leete's Island, 1980), 16–17.

2 For an account of this, see Jonathan Crary, *Techniques of the Observer: On Vision and Modernity in the Nineteenth Century* (Cambridge, MA: MIT Press, 1990).

3 William Henry Fox Talbot, *The Pencil of Nature* (New York: Da Capo Press, 1969), n.p.

4 William Henry Fox Talbot, 'Some Account of the Art of Photogenic Drawing, or, the Process by Which Natural Objects May be Made to Delineate Themselves without the Aid of the Artist's Pencil' (1839), in *Photography: Essays and Images*, ed. Beaumont Newhall (New York: Museum of Modern Art, 1980), 28. Emphasis in the original.

5 Louis Jacques Mandé Daguerre, 'Daguerreotype', in *Classic Essays on Photography*, ed. Trachtenberg, 12.

6 André Bazin, 'The Ontology of the Photographic Image', in *Classic Essays on Photography*, 241.

7 Roland Barthes, 'The Photographic Message,' in *Image-Music-Text*, ed. and trans. Stephen Heath (Glasgow: Fontana, 1977), 17.

8 The term 'indexical' is largely derived from the writings of the American logician C.S. Peirce. Within Peirce's basic triadic model of the sign (as index, icon and symbol), the index is directly connected to or affected by its object. This 'dynamical' or causal relation between the sign and its object distinguishes the index from both the icon (which only shares, through resemblance, the characteristics of the object it denotes, e.g. a portrait) and the symbol (which is an entirely constructed, and hence conventional, sign). However, these three aspects of the sign are not mutually exclusive and Peirce acknowledged, for example, that a photograph could be both indexical and iconic. For an introduction to Peirce's extensive writings on semiotics, see his 'Logic as Semiotic: The Theory of Signs', in *The Philosophy of Peirce: Selected Writings*, ed. J. Buckler (London: K. Paul, Trench, Trubner, 1940), 98–119. See also *Collected Papers of Charles Sanders Peirce*, 8 vols, ed. Charles Hartshorne and Paul Weiss (vols 1–6) and ed. Arthur Banks (vols 7–8) (Cambridge, MA: Belknap Press of Harvard University Press, 1931–58).

9 André Bazin, 'The Ontology of the Photographic Image', in *Classic Essays on Photography*, 241. Emphasis in the original.

10 Pierre Bourdieu, *Photography: A Middle-brow Art*, trans. Shaun Whiteside (Cambridge: Polity, 1990), 74. Emphasis in the original.

11 Joel Snyder, 'Documentary without Ontology', *Studies in Visual Communication* 10 (1984): 90.

12 Walter Benjamin, 'A Short History of Photography' (1931), in *Classic Essays on Photography*, 213. Emphasis in the original.

13 John Tagg, *The Burden of Representation: Essays on Photographies and Histories* (London: Macmillan, 1988), 160.

14 Roland Barthes, 'The Rhetoric of the Image', in *Image-Music-Text*, 39.

15 Allan Sekula, 'On the Invention of Photographic Meaning', in *Thinking Photography*, ed. Victor Burgin (London: Macmillan, 1982), 84–5. See also the chapters by Victor Burgin and John Tagg in this volume.

16 Bourdieu, *Photography: A Middle-brow Art*, 77.

17 Allan Sekula, 'On the Invention of Photographic Meaning', in *Thinking Photography*, ed. Burgin, 84.

18 Tagg, *The Burden of Representation*, 61. See also Allan Sekula, 'The Body and the Archive', in *The Contest of Meaning: Critical Histories of Photography*, ed. Richard Bolton (Cambridge MA: MIT Press, 1989), 343–88 (originally published in *October* 39 (1986)).

19 Abigail Solomon-Godeau, 'Who is Speaking Thus? Some Questions about Documentary', *Photography at the Dock* (Minneapolis, MN: University of Minnesota Press, 1994), 176. See also Martha Rosler's comments regarding 'victim photography' in her essay, 'In, Around, and Afterthoughts (on Documentary Photography)', in *Martha Rosler, 3 works* (Halifax, NS: The Press of the Nova Scotia College of Art and Design, 1981). An expanded version of this essay is published in *The Contest of Meaning*, ed. Bolton, 303–40.

20 *Night and Day: A Monthly Record of Christian Missions* (1 November 1877), 126. Emphasis in the original. For background on Barnardo, see Gillian Wagner and Valerie Lloyd, *The Camera and Dr. Barnardo* (London: National Portrait Gallery, 1974) and Gillian Wagner, *Barnardo* (London: Weidenfeld & Nicolson, 1979).

21 See especially John Szarkowski and Maria Morris Hambourg, *The Work of Atget*, 4 vols (New York: Museum of Modern Art, 1981).

22 Molly Nesbit, *Atget's Seven Albums* (New Haven, CT: Yale University Press, 1992), 16. Emphasis in the original.

23 Ibid.

24 Walker Evans, 'The Reappearance of Photography' (1931), in *Classic Essays on Photography*, 185.

25 Leslie Katz, 'Interview with Walker Evans', in *Photography in Print*, ed. Vicki Goldberg (Albuquerque, NM: University of New Mexico Press, 1981), 364. Emphasis in the original.

26 Quoted in William Stott, *Documentary Expression and Thirties America* (Chicago, IL: University of Chicago Press, 1973), 21.

27 Jeremy Bentham, 'Rationale of Judicial Evidence', in *The Works of Jeremy Bentham*, ed. John Bowring, vol. VII, Bk. IX, Pt. II, Ch. VI, Section 5 (Edinburgh: William Tait, 1843), 366–8.

28 John Grierson, 'The Documentary Idea: 1942', in *Grierson on Documentary*, ed. Forsyth Hardy (London: Faber & Faber, 1979), 112.

29 Stott, *Documentary Expression and Thirties America*, 8.

30 Quoted in Stott, *Documentary Expression and Thirties America*, 29. Emphasis in the original.

31 Elizabeth Cowie, 'The Spectacle of Actuality', in *Collecting Visible Evidence*, ed. Jane M. Gaines and Michael Renov (Minneapolis, MN: University of Minnesota Press, 1999), 19.

32 Lewis Hine, 'Social Photography', in *Classic Essays on Photography*, 111. All quotes from Hine are taken from this essay.

33 Maren Stange, *Symbols of Ideal Life: Social Documentary in America 1890–1950* (Cambridge: Cambridge University Press, 1989), 2.

34 Louis Jacques Mandé Daguerre, 'Daguerreotype', in *Classic Essays on Photography*, 12.

35 Quoted in Stange, *Symbols of Ideal Life*, 16

36 For a detailed critique of Riis and his ideological allegiances, see Sally Stein, 'Making Connections with the Camera: Photography and Social Mobility in the Career of Jacob Riis', *Afterimage* (May 1983): 9–15.

37 Hine, 'Social Photography', 111.

38 Beaumont Newhall, 'Lewis Hine', *Magazine of Art* 31, no. 11 (1938). Quoted in *Photo Story: Selected Letters and Correspondence of Lewis Hine*, ed. Daile Kaplan (Washington, DC: Smithsonian Institution Press, 1992), 105.

39 Beaumont Newhall, *The History of Photography: From 1839 to the Present* (New York: Museum of Modern Art, 1993), 235.

40 George Dimock, 'Children of the Mills: Re-Reading Lewis Hine's Child-Labour Photographs', *Oxford Art Journal* 16, no. 2 (1993): 52.

41 Edward Steichen, 'The F.S.A. Photographers', in *U.S. Camera 1939*, ed. T. J. Maloney (New York: William Morrow, 1938), 44.

42 Roy Emerson Stryker and Nancy Wood, *In This Proud Land: America, 1935–43, as Seen in the F.S.A. Photographs* (Greenwich, CT: The New York Graphic Society, 1973), 19. Recalling the Barnardo case, Lange was accused of fakery when she removed the mother's thumb from the lower right corner of the printed version of the image. Her FSA colleague, Arthur Rothstein, faced similar claims when it was revealed that he had placed, and then photographed, the same sun-bleached steer's skull against different backgrounds.

43 James Curtis, *Mind's Eye, Mind's Truth: FSA Photography Reconsidered* (Philadelphia, PA: Temple University Press, 1989), 52. See also Wendy Kozol, 'Madonnas of the Fields: Photography, Gender, and 1930s Farm Relief', *Genders* 2 (Summer 1989).

5 Interpreting vernacular photography: finding 'me' – a case study

by Catherine Whalen

What is vernacular photography? It may be easier to say what it is not. Vernacular photography often seems to mean non-art photography.[1] Yet, determining what is or is not art in this medium is hardly a simple matter, as debates dating from photography's early days to the present suggest. If one assumes that art museums confer the status of art upon objects displayed in their galleries, anonymous snapshots recently have become art photography, given the increasing number of exhibitions devoted to this popular form of cultural production.[2] Alternately, vernacular photography may signify amateur, as opposed to professional, photography. This distinction between untrained and formally trained practitioners undergirds the term 'vernacular architecture', where discussions of what is and is not vernacular antedate those taking place among historians of photography.[3] Yet here too, uncertainty about the import of the word 'vernacular' abounds.[4] One scholar of vernacular architecture, Bernard L. Herman, has described it as 'the architecture of common usage and communication'.[5] Here the emphasis is on 'vernacular' in the linguistic sense: the language shared by a group of people in a particular place during a period of time. Might, then, one also consider vernacular photography as the photography of common usage and communication? In which case, there are many vernacular photographies, found among various groups of makers *and* users across the globe since photography's inception. And, in which case, ubiquitous commercial genres such as portrait photography – produced by professional photographers in collaboration with their sitters on a staggeringly widespread scale – may claim vernacular status.[6]

If what matters in vernacular photography is social usage, approaches to studying it should address its 'communicative, performative and conversational qualities'.[7] To paraphrase historian Raphael Samuel's remarks on popular memory as unofficial history, vernacular photography constitutes a 'social form of knowledge; the work, in any given instance, of a thousand different hands'.[8] Photographs become comprehensible as they employ recognizable conventions, communicated through framing and focus, composition and cropping, posture and pose. At the same time, they resist intelligibility because of their inescapable ambiguity; inevitably, aspects of what they purport to show within their frames, or imply lies beyond them, are unknown and unknowable.

Commentators on snapshots in particular remark upon their simultaneous sameness and diversity, as makers repeatedly create images within familiar picture-making genres such as portraiture or landscape that nonetheless elude interpretation in the absence of more specific shared frames of reference. In spite of, or perhaps because of, this interplay of similarity and difference, the work of amateur photographers often arouses special interest. Why? In part, because so much photography takes place outside the professional spheres of art, documentary, photojournalism and commercial work. Billions of snapshots are made every year.[9] Moreover, they come to us in many guises, mediated by other forms of cultural production; they may appear attached to email messages or enclosed in letters, posted on websites or mounted in scrapbooks. In its plenitude and variety, amateur photography constitutes significant evidence of lived experience.

How might one interpret this evidence? It depends, of course, on the questions one seeks to answer. These photographs hardly lend themselves to a single mode of interpretation, especially in the absence of authorizing statements from known makers, curators or historians. Indeed, at least one commentator on discourses concerning found snapshots has expressed misgivings at how easily viewers attach their own narratives to them.[10] Such practices nonetheless confirm photography's vaunted capacity for polysemy. One might consider engaging these works, then, in ways that foreground multiple interpretations, and that address one's own relationship to them.

What follows is a case study offering one such approach to an example of amateur photography, an album made by a young woman in the United States during the 1920s.[11] As a form of evidence, the photographic album offers certain interpretive opportunities. First and foremost, it provides a purposefully selected set of photographs arranged in a particular sequence, which implies a narrative or story that unfolds page by page. This particular album also features captions, which enhance that narrative's legibility. In addition, the genre of the photographic album itself encodes a storytelling function. As Martha Langford has convincingly argued, these compendiums act as prompts for oral performance; when viewed in social settings, makers often explicitly narrate their albums while their audience looks and listens.[12] In the absence of those creators' narratives, how might one go about constructing others?

Finding 'me'

I first saw her while browsing in a second-hand store in Wilmington, Delaware. As I flipped through a sheaf of loosely bound pages pasted with photographs, there she was: a teenage girl perched demurely on an oak-and-leather settle (Fig. 5.1). In the overexposed print, sunlight from an unseen window dissolved one side of her face and body, and left the other in high relief. She wore a light-coloured, close-fitting dress with a lace bib-collar, a strand of pearls and a solemn expression. Underneath her photograph, she had simply written the caption 'me'; at the top of the page, the year '1924'. I closed the album, put it down and meandered on. As I rummaged through piles of old linens and stacks of mismatched china, her image stayed with me. I turned around, walked back and picked up the album again. It lacked a cover; instead, a cord fastened together leaves of heavy black paper. On them, 'me', whoever she was, had mounted pictures of herself, her family, her friends, her pets and her travels. With white ink, she carefully outlined the photos and inscribed captions. To some layouts, she added snippets of song lyrics such as 'Ain't we got fun?' and – more daringly – cigarette wrappers. Her pictorial narrative began in 1920 and ended some time after 1927.

Who was 'me'? From the pages of her photo album, she emerges as a consummate

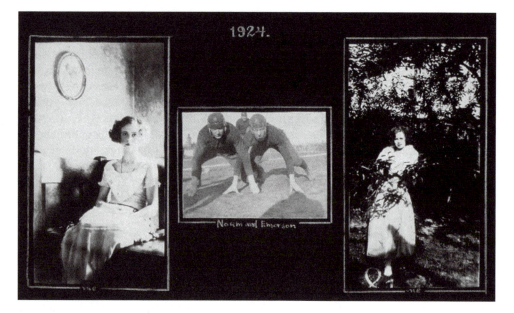

Figure 5.1 Mary von Rosen, 1924

American girl of the 1920s. A flapper and a flirt, she was white, middle-class and Mid-western. She loved a high-school football hero and adored horses. She lived outside Detroit, Michigan, but had family in Louisville, Kentucky. During her vacations, she cruised along the St Lawrence Seaway and visited a dude ranch in Wyoming. Towards the end of the album, pictures of her and a certain young man hint at a growing intimacy. Maybe she married him. With paper, photographs, scissors, glue and ink, she recorded and perhaps resolved her passage from girlhood to womanhood. 'Me' appears in 68 of the 232 photographs that she pasted into her album, but not once did she reveal her name. Of the nearly 200 people pictured in her album, she identified 46 by their first names and 25 by their full names. Among the latter individuals, only a handful emerge more than once, and none more often than her boyfriend 'Norm'. I wondered, if I found him, would I find her?

A Detroit newspaper clipping that she attached to her album's final page offered an important clue: an article profiling local high-school football captains, with a photo of Northern High's Norm Gabel. Here was the beau 'Norm' who, except for 'me' herself, appeared in the album more often than anyone else. Here, too, was a definite location: Northern High School in Detroit, Michigan.

In one of Norm's many pictures, he stands besides a new milkcar emblazoned 'Gabel's Creamery'. According to Detroit city directories, Philip Gabel, the owner of a creamery, lived within a few blocks of Northern High School during the early 1920s. I felt as though I was getting closer; maybe I would find out who 'me' was. But how? Perhaps a high school yearbook contained a photo of her that I could compare to those in her album. I called Northern High School and spoke to a school librarian. Yes, they had yearbooks from the early 1920s, and yes, they would look up alumni. She and a team of students found Norm Gabel, but not 'me'.

Maybe something or someone in Detroit might tell me who she was. So in the spring of 1999, I went there. With low expectations I too perused the yearbooks. In the January 1924 semi-annual, *The Viking*, I found Norman Gabel, just as he was pictured in the photo album. About 20 pages later, I found 'me', with a name beside her oval-framed image: Mary von Rosen. Her picture matched one in her album exactly (Fig. 5.2). Through her name, I recovered bits and pieces of Mary von Rosen's life history. Yet she told her tale best through her album. Without its painstakingly designed and subtitled images, her story would be lost. Mary von Rosen's photo album is what may be called a visual autobiography, straddling the distance between written self-works such as diaries, journals and letters, and pictorial ones like sketchbooks or scrapbooks.[13]

Figure 5.2 'When You and I Were Seventeen'

For authors constructing self-defining narratives, forms of life-writing invoke genres of drama as well as fiction. With their emphasis on genealogical time, and repetitive, ritualistic processes of creation, autobiographical works are re-enactments as well as novels. Analogously, the practices of amateur photography and album-making encompass both linear narratives and dramatic repertoires, as they consolidate multiple performances of identity into a single material artefacts. They are miniature theatres, complete with sets, costumes and props as well as stars and supporting casts. In other words, visual autobiographies distinctively concretize psychological space. Photograph albums house fantasy lives, accessible by invitation only. As individual creators produce these artefacts, they remake themselves as they work out ideal, alternative or potentially transgressive identities. Omissions are crucial. The process of forgetting – that is, editing the extraneous or the unwanted, first through photography and again through album-making – begets a remembered self.[14] Once it is complete, an album's narrative function achieves primacy. It becomes a record, destined to be replayed as a special chapter in one's life story.

How then, does the researcher or writer conceptualize his or her role in relation to self-works: as analyst, interpreter, storyteller or impresario?[15] Whichever posture a historian assumes, he or she is inevitably a voyeur and an informer. Gazing at one image after another in Mary von Rosen's album, I peer into her once-private life. Fingering the pages she carefully composed, I replay the intimate record that she left behind, one that she may have shared with only her closest companions. With its photographic vignettes set against black backgrounds, outlined and captioned in white, perhaps the album that Mary von Rosen fashioned resembles nothing so much as a silent movie. Directing the camera, she captured the actors, events and settings she deemed essential to her script. A skilled editor, she thoughtfully sequenced selected images and augmented them with the barest essential commentary. Details of her life outside the album constitute its back story, known to its original viewers but lost to chance twenty-first-century spectators. The task I have set for myself, then, has been to write a voiceover for a new audience, a public audience never meant to know her.

As a cultural historian, my first impulse was to sketch out a broad historical context. In 1920, the year Mary von Rosen began her album, women gained the right to vote, Prohibition went into effect and F. Scott Fitzgerald published *This Side of Paradise* and *Flappers and Philosophers*, immortalizing the so-called Jazz Age and its frankly sexual youth culture.[16] The ensuing decade witnessed repeated crises over ethnic and racial pluralism, and an increasingly entrenched consumer culture with a corporate ethos. But as I read book after book, article after article, recurring doubts about this interpretive strategy crystallized into a overriding conviction; Mary von Rosen's album was more than an archive from which to pluck illustrations for received historical narratives.

But how else might I grapple with it? As a scholar of visual and material culture, I then turned to the realm of amateur photography and album-making. With the advent and effective marketing of roll film and small hand cameras in the 1880s, such as the Kodak, amateur photography became popular throughout the United States; so did snapshot albums. Many of the amateur photographers who equipped themselves with easy-to-use, more affordable roll-film cameras were women, as manufacturers targeted middle-class female consumers who had both the disposable income and the leisure time to take up photography.[17] By the 1910s and 1920s, laced albums filled with sheets of blank paper like Mary von Rosen's were commonplace.[18] Magazines like *American Photography* offered amateurs explicit instructions for laying out their photo albums.[19] Also popular was a specialized form of memory book compiled by high school seniors. These commercially produced albums featured covers stamped with phrases like 'My Senior Year' or the gender-specific 'The Girl Graduate: Her Own Book'.[20] Conversely, high school yearbooks like Mary von Rosen's looked very much like photo albums.

But once again, I stopped short. Clearly, I could situate Mary von Rosen's album within any number of possible narratives, but which one? Ultimately, I turned to the narrative structure of the album itself. I followed its sequence, page by page, and wrote three distinct but overlapping accounts for each layout. First came 'her story'. As best I could, I described and analysed the composition of individual pages; inevitably, I hypothesized about their meanings. Then came the 'back story', in the details of Mary's life, once known to her album's contemp. obscure. Finally, there was 'my story', that is, my experiences as I hers. Here, too, was a place where I could freely speculate about Mary meanings, unencumbered by the burden of historical proof.[21] Of cours of these stories are mine; they constitute my album, in which I have fantasy realm that anyone who proposes to write history envisions for the

Mary von Rosen's album contains nearly a hundred pages of layout. one exception, the photographs of 'me' depict a teenager or a young woman. On the initial left-hand leaf, however, Mary showed herself as a child (Fig. 5.3).

Figure 5.3 Mary von Rosen as a child, Pingree Avenue

Her story

Three uncaptioned photographs form a row across the opening page of Mary von Rosen's album. Two snapshots show a young girl clutching a pair of dolls. A third image depicts the toys themselves. The backdrop for all three photos is the front-porch steps of a light-coloured brick house. In the central image, the girl sits on the top steps, the glass panes in the front door visible behind her. She is dressed in a heavy woollen coat and a felt hat with a brim. Her dark skirt shows below her coat hem, as do her knee socks and shiny shoes. One hand over the other, she hugs her doll to her chest and gazes wistfully up to the right. In another picture, she sits on the front lawn. She wears a white, short-sleeved dress trimmed with lace, and white socks and shoes. Her blonde hair is pulled back with a ribbon. Again she faces to her right, squinting in the sunlight, her toys perched on her lap. In the background, a woman rests on the steps.

Back story

The front porch steps are those of 231 Pingree Avenue in Detroit. The young girl in the photos is Mary von Rosen. She was born on 23 April 1907 in Illinois. Her parents, Ernest and Henrietta von Rosen, began their lives in Kentucky; so did her older sister Kate, born

904. Her younger brother, John, was born in Illinois in 1910. Soon thereafter, the von Rosens relocated to Detroit. By 1913, the family was living in this duplex at 231 Pingree Avenue.[22] Most likely, the photographs on the first page of Mary's album were taken shortly after the family moved into their new home. Mary would have been approximately six years old at the time, about the age she appears to be in these first few pictures.

Who were the von Rosens? A 1920 federal census-taker listed Ernest von Rosen as the head of the household. Her record identifies him as a 48-year-old white male, native-born of German immigrant parents. His 41-year-old wife Henrietta was also white and native-born. Her father was likewise a German immigrant, but her mother was born in Kentucky. According to the census, Ernest von Rosen was an 'expert accountant' in an auto factory.[23] As for religious affiliation, cemetery records reveal that the von Rosens were Roman Catholic.[24]

The von Rosen family moved to Detroit amid a huge influx of newcomers. Large-scale industrialization necessitated a vast labour force, and the city attracted one. During the 1910s, Detroit's population more than doubled. It reached nearly a million by 1920, making it the fourth most populous city in the United States.[25] The high wages offered by the rapidly expanding automobile industry made Detroit especially attractive.[26] As for Ernest von Rosen, city directories confirm that during the 1910s and early 1920s he held managerial positions with several different car companies, including Dodge Brothers.[27] The city's unprecedented growth strained its geographic and social boundaries. In the late nineteenth century, Detroit's neighbourhoods were typically ethnically rather than socio-economically homogeneous; there was a German section, a Polish one, and so forth. But in new communities like the von Rosen's, the residents tended to share the same race and class status rather than ethnicity.[28]

My story

The house, 231 Pingree Avenue, renumbered 897, still stands today. The front porch steps are identical to those that Mary von Rosen sat on in the photograph pasted to her album's first page. The door beyond bears the same glass panes. The facade is still clad with light-coloured brick. As I stood on the front sidewalk and snapped a picture with my camera, a taxi pulled up. A couple stepped out and walked toward the duplex. Somewhat suspiciously, the man called to me, 'You taking a picture of this house?' 'Yes,' I replied, adding, 'I know of someone who used to live there.' He looked surprised. 'Here? When?' I tell him, 'In the 1920s.' He broke into a smile and exclaimed 'The 1920s! That was a long time ago.' Our dialogue is not merely about time; it is also about race and place. In the 1920s, mostly native-born whites lived in this neighbourhood, whereas today primarily African Americans do. Only by invoking the past can I satisfactorily explain to him, a black man, why I, a white woman, photographed his home.

I compare my snapshot of 897 Pingree Avenue's front porch to those in Mary von Rosen's album. Mine looks slightly different: the paint is peeling, the viewpoint is higher. Undoubtedly Mary's photographer used a camera held against the chest rather than up to the eye. I wonder who took her picture, and if she photographed her dolls herself.

Leafing through her album, it becomes clear that Mary crafted her self-portrait not only by including certain kinds of images, but also by excluding others. Even though she presented herself as a child on her album's first page, she foregrounded herself alone, away from her family. Moreover, she chose images that depict her mothering toy infants, as if retrospectively affirming her own agency rather than her submission to choices that her parents made for her. Grouped together on a single sheet, these initial photographs constitute a portentous prologue to the narrative that begins in earnest on the following pages.

'Summer 1920 at Devil's Lake'

Her story

In a snapshot placed in the upper left corner, 'Me' sits decorously if incongruously atop a split-rail fence (Fig. 5.4). Now an adolescent, she wears a plain short-sleeved dress and her hair pinned up. Her eyes are downcast, her hands are folded in her lap, and her ankles, not her knees, are crossed. Photographs strung diagonally from lower left to upper right feature well-dressed young adults beside a lake: Helen; Pete; Bob and Helen. In one picture, Bob and Helen stand side by side, hands entwined; clearly they are a couple. In another, Pete, a handsome young man in a coat and tie, casually poses solo with his hands in his pockets. Below his image is a cut-out of Mary, sporting a close-fitting, one-piece bathing suit and a cascade of long blonde hair. She stands with her feet together, arms thrust back and torso arched forward as if readying for a dive.

Figure 5.4 'Summer 1920 at Devil's Lake'

Back story

Mary was 13 years old when she visited Devil's Lake. Michigan abounds with lakes of that name; this one may be in Genesee County, perhaps 60 miles northwest of Detroit. Travelling by car, it would have been an easy trip from the von Rosen's new home at 867 Edison Avenue. Their two-storey, wood-frame single-family house was seven blocks north of their former Pingree Avenue duplex and just within the confines of the more fashionable Boston-Edison neighbourhood.[29] Perhaps the entire von Rosen family took a vacation together, but Mary chose to feature only selected players in her portrayal of summer 1920 at Devil's Lake.

That summer, women's swimwear made the headlines. Municipalities across the United States were just beginning to lift prohibitions against women wearing 'scanty one-piece bathing suits' at public beaches.[30] Banned or not, bare-limbed bathing beauties epitomized female attractiveness. Embodied in much-admired women swimming champions and their movie-screen manifestations, these icons cloaked their eroticism in wholesome athleticism. In her photograph, Mary von Rosen claimed her place among them.[31]

My story

On this single page Mary lays out some of her album's major themes. Foremost among them is her burgeoning sexuality. In this sequence's first image, her pose and attire mirror that of her childhood self on the previous page. Given the scrapbook's inherent physical chronology, six-year-old Mary faces to her right, as though looking back at her past, while her adolescent incarnation inclines her gaze downward to her left, as if contemplating the future but not yet facing it. Helen, first pictured alone, then together with Bob, embodies a staple romantic fantasy: courtship and couplehood. Pete signals one of Mary's recurrent preoccupations: boys. In her second picture, Mary displays her developing body for the camera, looking directly into the lens. She has discovered not only her physical beauty, but also its strategic deployment.

'Summer 1921 at Belle Isle'

Her story

The next instalment, 'SUMMER 1921', expands to two leaves. First come an array of dapper young men in suits and ties with slicked-back, centre-parted hair: Paul Livingston, Cecil Brown and Cub Parker, plus another simply identified as 'Smith'. In one photograph, Mary stands besides Paul, brightly beaming and smartly dressed in a dark slim skirt, a bolero jacket and a ruffled blouse.

The second page features a snapshot of a young child named Anne standing on a car's running board, one photo of Mary, and three pictures of Billy, a teenage boy (Fig. 5.5). Among the latter images is a close-up of Billy's head and upper body that Mary cropped into the shape of an octagon. In another shot, he poses on a beach in a bathing suit. In the third photo he stands beside a canoe underneath a willow tree on the shores of a lake. Clad in white pants and a white shirt with the sleeves rolled up, he thrusts his hands into his pockets. The bottom corner of his picture extends over the upper corner of Mary's photograph. Sitting demurely on the prow of the same canoe, she wears a dark-coloured dress with a white lace collar, and black shoes and stockings. She clasps her hands in her lap and smiles for the camera, or Billy, or both. At the juncture where the two images overlap, Mary wrote 'Us – at Belle Isle'.

Figure 5.5 'Summer 1921 at Belle Isle'

Back story

In the spring of 1921, Mary completed her first year at Northern High School. That summer, she and her fellow Detroiters flocked to nearby Belle Isle, a 1000 acre island in the Detroit River. They picnicked in the park, canoed in the canals, strolled along the promenade, and admired the fountains and gardens. When the city of Detroit originally purchased Belle Isle as a public park in 1879, naysayers decried it as a municipally sponsored pleasure ground for the rich.[32] Soon, however, Belle Isle's supporters dubbed it the 'People's Park',[33] and hailed it as a much-needed restorative for the general public.

My story

Mary projects a mixture of sophistication and vulnerability on these pages, while her parents and siblings remain glaringly absent from them. Her sister Kate received her diploma from Northern High School in 1921, but no graduation picture of her appears in Mary's album.[34] Here and elsewhere, Mary consistently elides family rituals; instead, she indexes her prospective beaux. As she asserts her popularity through collecting snapshots of her many presumed admirers, she also takes the first tentative steps towards couple-hood with one of them. Yet the overlapping photos of 'Us – at Belle Isle' connote an illusory intimacy. Despite their retroactive conjoining, the photographs of 'Billy' and 'Me' remain distinct images, each self-contained and labelled with individual captions as well as a tentative shared one.

'Summer 1922 at Pine Lake'

Her story

A few pages later, 15-year-old Mary commemorated the summer of 1922 with her most exotic handiwork, producing a two-page spread devoted to revelry at Pine Lake (Fig 5.6). The left leaf sports a collage of photos and mementos inscribed 'That CRAZY Summer 1922'. Among them is a pink ribbon emblazoned 'Souvenir of the FAIR – HAVING A GOOD TIME'. Nestled underneath a red bow is a miniature oar on which Mary inked 'House Party at Pine Lake'. In between red feathers she interspersed cut-outs of two teenage girls in bathing suits, Kate and Jean. Mary adorned the base of a fan-shaped arrangement of cigarette wrappers with a heart cut out of black paper and a diagonal of red cord. Alongside a picture of Jean dressed in jacket and skirt she placed a cigarette full of tobacco. The opposite page, sedately titled 'SUMMER 1922 AT PINE LAKE', features five photographs arranged more symmetrically. Within them, however, the hoopla continues. A cut-out of Lillian Hoff, an oval vignette of Katie and Jean in jodhpurs, a circular medallion of Harry Martin, and another shot of Katie Besancon flank a central image captioned 'Ain't we got fun?' A faded photograph reveals a group of boys and girls in bathing suits posed on a lawn. Among the boys kneeling in front, one youngster appears as if he might break into song, one hand on his upraised knee and the other on his heart. Behind them stand a row of girls, a bobbed-haired Mary among them. Small lakeside cottages appear in the backdrops of several photos.

Figure 5.6 'That Crazy Summer 1922' / 'Summer 1922 at Pine Lake'

Back story

During 'That CRAZY Summer 1922', Mary von Rosen and her pals most likely frolicked at Pine Lake in nearby Oakland County, perhaps 15 miles northwest of Detroit. Of all the friends or acquaintances who took part in the 'House Party at Pine Lake', Katie Besancon proved the most long-standing. That summer, Katie was 17, two years older than Mary. A 1922 chronicle of Detroit luminaries reveals that she was the daughter of William E. Besancon, who owned and operated one of the city's largest and most successful coal yards.[35] City directories from the early 1920s disclose that the Besancons resided across the street from the von Rosens. They made their home in a two-storey, brick-veneered wood-frame house at 932 Edison Avenue.[36] In it lived Katie, her father, her mother Sybilla, her brother William, Jr, and a servant named Lizzie Schwass, a German immigrant.[37] Both William Besancon and his wife were native to Detroit, although his family were the more recent arrivals; his mother and father emigrated from France with their respective families in the mid-1800s. The Besancons were Episcopalian, and unlike Mary, Katie attended a private rather than public high school.[38]

In the summer of 1922, Detroiters like Mary and Katie may have had a special fondness for the recent hit song *Ain't We Got Fun*. Richard Whiting, once employed in the Detroit office of eminently successful song publisher Jerome H. Remick, composed its music, while Chicagoan Gus Kahn and Raymond Egan, a former Detroit bank clerk, wrote the words. The 1921 tune's catchy, danceable melody and devil-may-care lyrics enjoyed widespread popularity. Despite, or perhaps because of, its refrain's disingenuous glee in the face of financial adversity – 'Not much money,/Oh! But honey, Ain't we got fun' – the song reaped great profits, selling over a million copies.[39]

Exemplified by the lyrics of *Ain't We Got Fun*, early 1920s American popular culture – stories, songs, motion pictures, advertisements, magazine features and newspaper articles – explicitly celebrated the pursuit of pleasure. More often than not, however, such enjoyment entailed procuring a newly desirable good or service. Driving the shifts in early-twentieth-century manners and morals was the dynamic interrelationship of booming consumerism, individual subjectivity and social behaviour played out along the axes of race, nationality, class, gender, age and place. Among the white middle class, would-be sophisticates reinscribed once private, upper-class leisure-time pursuits and more public working-class amusements within a broader circle of bourgeois acceptability, if not respectability. At the same time, they reiterated their class status as well as national and racial identities through the places they lived, the jobs they held, the people they married, the goods they purchased, and those they distanced themselves from.

Teenage girls like Mary von Rosen fashioned their self-presentations within a peer-driven youth culture that idolized the flapper; that is, 'the nice girl who is a little fast'.[40] Flippant and vivacious, the model flapper was a fun-loving young woman whose sex appeal and air of indifference towards men attracted hordes of lovelorn swains, with whom she mercilessly flirted. As many commentators have noted, then and later, flappers emulated heretofore masculine behaviour: they cut their hair short, revealed their legs, used slang, drank and smoked. Through consumption, these fashionable young women reduced certain differences between gender roles and asserted their symbolic if not actual equality with men. At the same time they opened themselves up to new modes of sexual objectification and exploitation.[41]

Whether or not Mary and her girlfriends possessed ideal flapper attributes, each could pass as one by purchasing key props and employing them in prescribed performances. Chief among such demonstrations was smoking cigarettes. Once women who smoked proclaimed themselves bohemians, intellectuals or prostitutes, but after the First World War, cigarette manufacturers began to market their product to young people and especially women. Advertisers and trend-setting female college students alike avowed a woman's right to smoke publicly, while traditionalists who held white middle-class women responsible for maintaining the country's moral and social standards feared one lapse in propriety would soon lead to others.[42]

My story

Cigarettes dominate Mary's collage of the culture of pleasure. Their potent symbolism did not escape her deployment, at least in the private world of her album. What if Mary's mother encountered her daughter's handiwork? Would she have deemed the Chesterfield and Lucky Strike wrappings, or the tobacco-filled 'SINBAD Special', evidence of a particularly daring transgression? The cigarette paraphernalia's juxtaposition with photographs of Katie and Jean imply that Mary was more concerned about her peer's perception of her smoking than her parents' possible disapproval of her apparent behaviour. If anyone besides Mary ever scrutinized her album, presumably her girlfriends did.

For Mary von Rosen, an idol more accessible than the mythic flapper lived right across the street: Katie Besancon. A couple of years older and from a wealthier, more established family, she may have embodied all that Mary wished to be. In time Katie would marry a prominent Detroit lawyer and move to Gross Pointe Farms, an exclusive suburb.[43] Mary might have fantasized about a similar future, but her trajectory towards it would not have been as self-evident.

A 9 June 1963 newspaper clipping from the *Detroit Free Press* reveals that Katie Besancon, long since Mrs Frank Boos, spent three years creating a lavishly furnished doll's house in anticipation of a grandchild who turned out to be a boy. Looking at the photograph of her miniature masterpiece and reading the description of her custom-made furniture replicas, I pondered the precise, painstaking construction of Mary's album. Perhaps Katie and Mary made albums in tandem, each surveying, admiring and critiquing the other's handiwork as they transposed their respective histories and fantasies into material narratives.

Conclusion

For Mary von Rosen, that narrative continued for another 70 or so pages, as she selected, arranged, affixed and captioned this visual chronicle of her transition from adolescence to adulthood. Through my tripartite mode of interpretation, I add to her account, conjuring up 'me' past and present. With this approach I seek to honour her album's formal qualities and affective power. I also explore how, through her manipulation of the photo album genre, she imagined possible life scripts within her milieu, and how her material interpolation of self and culture may illuminate historical understanding of this place and time.[44] Considering the album through multiple frames allows for conflicting accounts, revealing possible divergences between idealized and lived experience. My explicit self-positioning highlights the constructedness of my interpretations, especially their speculative, contingent aspects. In foregrounding Mary's agency as well as my own, I posit us as collaborative narrators; between us, we have still more stories to tell.

This case study begins to suggest the challenges of interpreting vernacular photography. What modes of address are appropriate for these kinds of objects? I have suggested three possible interpretive frames, entailing visual and textual analysis, historical contextualization and self-reflexivity; what others might prove productive? One might consider how vernacular photography circulates, through what forms of exchange; or its various repositories, whether archive or trash bin, virtual or otherwise. Undoubtedly there is a growing market for more antiquated forms of vernacular photography, especially those that predate digital images.[45] In part, collecting drives scholarship on this topic, as happens in many areas of object study; in turn, collecting practices become objects of study in their own right. To return to the example of snapshot shows in art museums, several feature an individual's collection, including some recently gifted to the exhibiting institutions and documented through accompanying catalogues.[46] The common point of reference for these photographs is necessarily the private collector, and subsequently, the collecting institution.

If social usage is the hallmark of vernacular photography, then considering how it is embedded in human relationships is crucial, whichever matrices of relationships one chooses to explore. Likewise important is grappling with the materiality of these forms of cultural production, or the interplay of their physical aspects and social life.[47] In addition, the volume and ubiquity of vernacular photography, coupled with its specificity and intimacy, directs attention to the relationship between the global and the local.[48] How does one negotiate the generic and the particular, or familiarity and strangeness? Such dilemmas raises further questions about how one apprehends forms or practices of vernacular photography beyond one's own perception of what is customary, exposing how terms such as 'common', 'ordinary' or 'everyday' must be used carefully lest they elide difference and mask power relations.[49] Nevertheless, if vernacular photography is the work of billions of different hands, so too is its study. In which case, we all have many stories to tell.

Notes

1   Geoffrey Batchen, 'Vernacular Photographies', in *Each Wild Idea: Writing, Photography, History* (Cambridge, MA: MIT Press, 2001); Geoffrey Batchen, 'Snapshots: Art History and the Ethnographic Turn', *Photographies* 1, no. 2 (2008): 121–42. For a historiography and bibliography of studies in vernacular photography, see Stacey McCarroll Cutshaw and Ross Barrett, 'In the Vernacular: Photography of the Everyday' and 'Bibliography', in Cutshaw and Barrett, *In the Vernacular: Photography of the Everyday* (Boston, MA: Boston University Art Gallery; Seattle, WA: University of Washington Press, 2008).

2   In many such exhibitions, the selection of snapshots according to some aesthetic criteria, whether by curators or collectors, becomes paramount. See Douglas R. Nickel, *Snapshot: The Photography of Everyday Life, 1888 to the Present* (San Francisco, CA: San Francisco Museum of Modern Art, 1998); Mia Fineman, ed., *Other Pictures: Anonymous Photographs from the Thomas Walther Collection* (Santa Fe, NM: Twin Palms, 2000); *Close to Home: An*

American Album, intro. D.J. Waldie (Los Angeles, CA: J. Paul Getty Museum, 2004); Barbara Levine et al., *Snapshot Chronicles: Inventing the American Photo Album* (New York: Princeton Architectural Press; Portland, OR: Douglas F. Cooley Memorial Art Gallery, 2006); Sarah Greenough et al., *The Art of the American Snapshot, 1888–1978: From the Collection of Robert E. Jackson* (Washington, DC: National Gallery of Art; Princeton, NJ: Princeton University Press, 2007); Marvin Heiferman et al., *Now is Then: Snapshots from the Maresca Collection* (New York: Princeton Architectural Press; Newark, NJ: Newark Museum, 2008).

3   For a discussion of studies of vernacular architecture and vernacular photography, and a thoughtful interrogation of the term vernacular, see Bernard L. Herman, 'Vernacular Trouble: Exclusive Practices on the Margins of an Inclusive Art', in Cutshaw and Barrett, *In the Vernacular*.

4   Dell Upton, 'The Power of Things: Recent Studies in American Vernacular Architecture', in *Material Culture: A Research Guide*, ed. Thomas J. Schlereth (Lawrence, KS: University Press of Kansas, 1985), 57–9; Herman, 'Vernacular Trouble'.

5   Bernard L. Herman, *Architecture and Rural Life in Central Delaware, 1700–1900* (Knoxville, TN: University of Tennessee Press, 1987), 13.

6   Many photographs designated vernacular are portraits, produced by both professional and amateur photographers, and in some cases further embellished by their makers and users; they include daguerreotypes, ambrotypes, tintypes, albumen prints and snapshots. See Batchen, 'Vernacular Photographies'; Batchen, *Forget Me Not: Photography and Remembrance* (Amsterdam: Van Gogh Museum; New York: Princeton Architectural Press, 2004). Likewise, given their extensive distribution, one may consider advertising, news, souvenir, and erotic photographs to be examples of vernacular photography. See Cutshaw, preface to *In the Vernacular*, 7–8.

7   Herman, 'Vernacular Trouble'.

8   Raphael Samuel, *Theatres of Memory* (London: Verso, 1994), 8.

9   Snapshots date from the introduction of the Kodak camera in 1888; the term came into popular usage from hunting, where it denoted a quick, loosely aimed shot. Since the late 1960s, billions of snapshots have been taken annually in the United States. By recent count, over two billion images are currently available via Flickr alone, one of the most popular photo-sharing websites, even as its contributors blur boundaries between amateur and professional practice. See Sarah Greenough, introduction to *The Art of the American Snapshot*, 2, 4, 6; and Virginia Heffernan, 'Sepia No More', *New York Times*, 27 April 2008.

10   Nancy West, 'Telling Time: Found Photographs and the Stories They Inspire', in Heiferman et al., *Now is Then*. West insightfully identified three prominent narratives surrounding found photographs, those of discovery, accident and oblivion.

11   An abbreviated version of this case study appeared in a special issue of *Afterimage* devoted to vernacular photography; see Catherine Whalen, 'Finding "Me"', *Afterimage* 29, no. 6 (2002): 16–17. I am indebted to Glenda Gilmore, Laura Wexler, Dolores Hayden, Nancy Cott and members of the Photo Memory Workshop at Yale University for their thoughtful comments on this project.

12   Martha Langford, *Suspended Conversations: The Afterlife of Memory in Photographic Albums* (Montreal: McGill-Queen's University Press, 2001).

13   See Marilyn F. Motz, 'Visual Autobiography: Photograph Albums of Turn-of-the-Century Midwestern Women', *American Quarterly* 41, no. 1 (1989): 63–92; Susan Brynteson and L. Rebecca Johnson Melvin, *Self Works: Diaries, Scrapbooks and Other Autobiographical Efforts* (Newark, DE: University of Delaware Library, 1997); Katherine Ott, Susan Tucker and

Patricia Buckler, 'An Introduction to the History of Scrapbooks', in *The Scrapbook in American Life*, ed. Susan Tucker, Katherine Ott and Patricia Buckler (Philadelphia, PA: Temple University Press), 1–25.

14 On photography and memory, see Barbie Zelizer, *Remembering to Forget: Holocaust Memory through the Camera's Eye* (Chicago, IL: University of Chicago Press, 1998).

15 Margaret Atwood, 'AHR Forum: Histories and Historical Fiction/In Search of Alias Grace: On Writing Canadian Historical Fiction', *American Historical Review* 103, no. 5 (1998): 1503–16; John Demos, 'AHR Forum: Histories and Historical Fiction/In Search of Reasons for Historians to Read Novels . . .', *American Historical Review* 103, no. 5 (1998): 1526–9; Lynn Hunt, 'AHR Forum: Histories and Historical Fiction/"No Longer an Evenly Flowing River": Time, History, and the Novel', *American Historical Review* 103, no. 5 (1998): 1517–21; Jonathan D. Spence, 'AHR Forum: Histories and Historical Fiction/Margaret Atwood and the Edges of History', *American Historical Review* 103, no. 5 (1998): 1522–5; Lorna Marie Irvine, 'Writing Women's Lives: My Self, Her Self, Our Selves', *American Review of Canadian Studies* 24, no. 2 (1994): 229–40; Motz, 'Visual Autobiography'.

16 F. Scott Fitzgerald, *This Side of Paradise* (New York: C. Scribner's Sons, 1920) and *Flappers and Philosophers* (New York: C. Scribner's Sons, 1920).

17 Colin Ford, ed., *The Kodak Museum: The Story of Popular Photography* (Bradford, UK: National Museum of Photography, Film and Television, 1989), 60–1, 65, 67.

18 Richard W. Horton, 'Photo Album Structures, 1850–1960', *Guild of Book Workers Journal* 32, no. 1 (1994): 32–43.

19 E.A.E. Stewart, 'Improving the Album', *American Photography* 10, no. 12 (1969): 672–5.

20 See 'Susan Tucker, Works in Progress', www.tulane.edu/~wclib/archive.html for 'My Senior Year', owned by Helen Lowe, collection of the Nadine Vorhoff Library, Newcomb College Center for Research on Women; and Brynteson and Melvin, *Self Works*, 27, for 'The Girl Graduate: Her Own Book', published by Reilly and Britton in Chicago, owned by Grace Lloyd and dated June 1913, Special Collections, University of Delaware Library.

21 Informing this approach is the method of artefact analysis developed by Jules Prown, especially his process of description, deduction and speculation. See Jules David Prown, 'Mind in Matter: An Introduction to Material Culture Theory and Method', *Winterthur Portfolio* 17, no. 1 (1982): 1–19.

22 The 1913 Detroit city directory.

23 The 1920 federal census for Detroit, Wayne County, Michigan.

24 Von Rosen family plot records, Holy Sepulchre Cemetery, Southfield, Michigan.

25 Sidney Glazer, *Detroit: A Study in Urban Development* (New York: Bookman Associates, 1965), 79, 94.

26 In 1908, the city's automotive manufacturers employed 7200 workers; in 1916, they employed 120,000. In 1919, their combined labour forces encompassed 45 per cent of Detroit's industrial workers. Between 1913 and 1920 the average daily wage increased from $2.60 to $6.20, propelled in part by Henry Ford's 5 dollars per day wage rate, which he established in 1914 and increased to 6 dollars per day in 1919. Notably, Ford excluded certain employees from the much-publicized wage rate, among them probationary workers, most women, unmarried men under 22 years old and divorcees. See Glazer *Detroit*, 79, 94; Melvin G. Holli, ed., *Detroit* (New York: Near Viewpoints, 1976), 119; Robert Shogan and Tom Craig, *The Detroit Race Riot: A Study in Violence* (Philadelphia, PA: Chilton, 1964), 19; B.J. Widick, *Detroit: City of Race and Class Violence* (Chicago, IL: Quadrangle, 1972), 33.

27 The 1914–17 and 1920–21 Detroit city directories. The Briscoe Motor Corporation (1914–21) was based in Jackson, Michigan, while Detroit was home to the Saxon Motor Car Corporation (1914–22) and Dodge Brothers (est. 1914; merged with Chrysler in 1928). See Tad Burness, *Cars of the Early Twenties* (Philadelphia, PA: Chilton, 1968), 3, 5, 10.

28 Oliver Zunz, *The Changing Face of Inequality: Urbanization, Industrial Development, and Immigrants in Detroit, 1880–1920* (Chicago, IL: University of Chicago Press, 1982), 3, 130–52.

29 The 1920–21 Detroit city directory.

30 C.1920 newspaper clippings, Zeta Collection, Seeley G. Mudd Library, Yale University.

31 For a discussion of the first Miss America contest in Atlantic City, New Jersey in 1921 vis-à-vis the popularity of Australian swimming champion Annette Kellerman and filmmaker Mark Sennett's bathing beauties, see Lois W. Banner, *American Beauty: A Social History through Two Centuries of the American Idea, Ideal and Image of the Beautiful Woman* (New York: Knopf, 1983), 265–70.

32 Frank Barcus, *All around Detroit* (Detroit, MI: Frank Barcus Art Studio, 1939), 67.

33 Wilma Wood Henrickson, ed., *Detroit Perspectives: Crossroads and Turning Points* (Detroit, MI: Wayne State University Press, 1991), 225.

34 *The Viking* 4, no. 8 (1921): 38. In 1921, the students of Northern High School, Detroit, Michigan, published *The Viking* monthly. On p. 38 of the June issue, Kate von Rosen's photograph appears among those of the graduating senior class.

35 Clarence Monroe Burton, *The City of Detroit, Michigan, 1701–1922*, vol. 5 (Chicago, IL: S.J. Clarke, 1922), 558.

36 The 1920–21 Detroit city directory.

37 The 1920 federal census for Detroit, Wayne County, Michigan.

38 Burton, *City of Detroit*, 558.

39 Gus Kahn, Raymond B. Egan and Richard A. Whiting, 'Ain't We Got Fun', in *Ain't We Got Fun: The Great Songs of Richard Whiting* (Secaucus, NJ: Warner Bros., 1991), 10–12; David Jasen, *Tin Pan Alley* (New York: Donald I. Fine, 1988), 30, 122.

40 Elizabeth Stevenson, 'New Social Types', in *Ain't We Got Fun? Essays, Lyrics, and Stories of the Twenties*, ed. Barbara H. Solomon (New York: New American Library, 1980), 236.

41 See Paula S. Fass, *The Damned and the Beautiful: American Youth in the 1920's* (New York: Oxford University Press, 1977).

42 John C. Burnham, *Bad Habits: Drinking, Smoking, Taking Drugs, Gambling, Sexual Misbehavior, and Swearing in American History* (New York: New York University Press, 1993), 96–7.

43 'Mrs. Frank Boos, Widow of Attorney', *Detroit News*, 8 October 1976.

44 On life scripts, see Carolyn G. Heilbrun, *Writing a Woman's Life* (New York: W.W. Norton, 1988).

45 See Samuel, 'The Discovery of Old Photographs', in *Theatres of Memory*, 337–49; Cutshaw and Barrett, 'In the Vernacular', 24; Paul Grainge, *Monochrome Memories: Nostalgia and Style in Retro America* (Westport, CT: Praeger, 2002).

46 Greenough et al., *The Art of the American Snapshot*; Heiferman et al., *Now is Then*.

47 My understanding of materiality is informed by Bernard L. Herman, 'Troublesome Things: The Dwelling, The Teapot, the Quilt, the Oyster', *William and Mary Quarterly*, forthcoming.

48 See Christopher Pinney and Nicolas Peterson, *Photography's Other Histories* (Durham, NC: Duke University Press, 2003); Batchen, 'Snapshots: Art History and the Ethnographic Turn'.

49 Dell Upton, 'Architecture in Everyday Life', *New Literary History* 33, no. 4 (2002): 707–23; Herman, 'Vernacular Trouble'.

6   Newsreels: form and function

by Luke McKernan

A newsreel *was* a reel of film showing a collection of news stories released at regular intervals in cinemas. That is a definition for use with the past tense. Alternatively one could say that a newsreel *is* a reel of film containing reports of past events, found in archives and utilized chiefly by television programmes seeking to illustrate historical events. The newsreel enjoyed a production and exhibition history for much of the twentieth century, and has since enjoyed an afterlife in which its contents can be repackaged as forms of historical illustration, or evidence. Yet the newsreel also illustrates that evidence is never absolute, as critics challenge the newsreel's claim to display meaningful evidence of any kind, and television audiences turn away from monochrome visions of the past to the greater visual appeal of dramatized historical sequences and the charms of computer-generated imagery. What the newsreel *will be* in the future, for academic researchers, television viewers, Internet users and audiences as yet unknown, must be uncertain. What is required is a sympathetic understanding of the forms which the newsreel took, including its interrelationships with other media. To appreciate the newsreels for the evidence they might provide, it is important not to see them in isolation.

Newsreels were a creation of the cinema. News on film had existed since 1895, when the British film pioneers Birt Acres and Robert Paul produced fleeting reports of the Oxford–Cambridge Boat Race and the Epsom Derby for exhibition in the peepshow Kinetoscope viewer. Other such news reports became common in the first ten years or so of film exhibition, with particular emphasis on sporting events, ceremonial occasions and war reportage (notably the Anglo-Boer War, Spanish-American War and Russo-Japanese War). However, such films were exhibited irregularly in music halls and variety theatres. They lacked a fixed audience. When cinemas of various kinds began to emerge in the early twentieth century, with the general solidifying of the motion picture exhibition business, people adopted regular habits of attendance, usually once or twice a week. The audience was fed a consistently structured programme of films, with the knowledge that different titles, nevertheless offering up more of the same, would be available to them at their next visit. News thrives on such regularity, because what makes the news is not simply the content of the news medium but the expectations of a particular audience. News is defined by the locality, outlook and understanding of its specific consumers; what is news to one audience may not be news to another. Cinemas created a particular audience clientele; they created the newsreels.

The first newsreel is generally said to have been *Pathé Fait-Divers*, initially exhibited in Paris in 1908 and then across France, when it became *Pathé Journal* in 1909. Pathé established the newsreel form from the outset. The reel brought together disparate stories, some national, some international (a strength of the multinational Pathé organization), all linked by a shared topicality, and released weekly. Newsreels from its French rivals Gaumont, Eclair and Eclipse soon followed, and Pathé exported the model overseas. The first British newsreel was *Pathé's Animated Gazette*, established in 1910, and rapidly followed by *Warwick Bioscope Chronicle* (1910), *Gaumont Graphic* (1910) and *Topical Budget* (1911). In the United States, *Pathé's Weekly* appeared in 1911, and was soon joined by such indigenous offerings as *Vitagraph Monthly of Current Events* (1911), *Mutual Weekly* (1912) and *Universal Animated Weekly* (1913).

The periodicity of such newsreels was largely dependent on geography. In the United States, newsreels came to be issued weekly; in smaller Britain, the norm was bi-weekly. Each newsreel could be expected to be shown in cinemas daily until supplanted by the next issue, but many cinemas showed older newsreels, which were priced according to their degree of freshness. Newsreels in Britain could have a shelf life of six weeks or more, which meant that at any one time cinemas might be showing newsreels of a variety of ages, in a complex maze of distribution, according to the prices cinemas were prepared to pay for them. Consequently, the newsreels often favoured stories that were not particularly tied to a specific date, such as novelty items, scenic stories or items which privileged a striking image of some kind, because these best fitted a medium that had to appear relevant over a period of weeks, as well as being up-to-the-minute where it could.[1] This was news that was both as immediate as it could be and effectively timeless, or time-free.

Newsreels sprang up at the same time in other countries. By the time of the First World War, Germany had *Tag im Film* (1911), *Union-Woche* (1913), *Eiko-Woche* (1913) and *Messter-Woche* (1914). Russia had *Mirror of the World* (produced by Pathé) and Australia had *Australasian Gazette*; many others relied on issues imported from Pathé or Gaumont, with the occasional locally shot story inserted. The internationality of the medium, with international exchange of stories proving cheaper than basing camera operators overseas, encouraged conformity of content and style. The reel would be introduced by a main title trumpeting the newsreel itself, then each story (at this period silent, of course) would be introduced by a main title and sometime subtitle, in the style of newspaper headlines, before the newsfilm itself followed, generally at this time uninterrupted by any further titles. The filming style itself was basic. Generally one camera operator was used per story, and events were depicted through a plain accumulation of four or five shots denoting the main visual points of the action. The emphasis was always on the visual; this was news to be seen. The sensational and the popular inevitably predominated over the analytical. The newsreels also took their cue from stories which had been established by other media, namely newspapers, particularly the illustrated papers. Newsreels inevitably were released some days after the events that they depicted. This put them at the end of the news chain, but it also gave them their particular role. They provided the pictures in animation of stories that were known to their audience, either through familiarity (annual ceremonies, common sporting events etc.) or because they had been established as news by the newspapers. The newsreels showed audiences what they had been discussing; they completed the picture.

The First World War saw the newsreel established as a common feature of the conventional cinema programme. It also saw the first interest in the medium from governments, which had to take notice of the millions now drawn to the cinema screen on a regular basis. The French created *Annales de la Guerre*, the British War Office took over an existing newsreel and formed *War Office Official Topical Budget*, later *Pictorial News (Official)*, while in the United States the Committee on War Information released *Official War Review*. Such newsreels benefited from exclusive access to war front footage through systems of officially recognized camera operators, but in tone and style were no more overtly propagandist than the commercial newsreels, as they had to overcome resistance from a film industry suspicious of government-supported productions, which they saw as having an unfair advantage. A propaganda newsreel was an attractive proposition for a government in theory, but the promise of news management had to be balanced by the need to put out a product which could hold its own in the marketplace.

The 1920s was the first great era of the newsreels, when the social, sporting and political whirl of the era found expression in the rapid pot-pourri of the newsreel style. John Dos Passos, in his *U.S.A.* novel trilogy, called his headline sections 'Newsreel', and a newsreel technique is employed to portray the colourful, dynamic and visual nature of modern American society. The newsreel had become a quintessential American medium, and it was the Hollywood studios which were to become dominant as the parent companies of the newsreels as they adopted sound at the end of the decade and enjoyed their great period of dominance and popularity in the 1930s. Fox Movietone, Paramount and Universal became the titles most familiar to newsreel viewers around the world, even if the names of Pathé and Gaumont (sometimes with little connection remaining with the original French companies) were still widespread. Some countries managed to support modest local reels, such as Polygoon's *Hollands Nieuws* in the Netherlands, but in general a newsreel had to have international reach, giving the multinationals an unassailable advantage. Notable also for the future development of the newsreel was the situation in Italy, where Mussolini's Fascist government from 1926 ran the Luce Institute, including a nationalist newsreel which exhibitors were obliged to screen at every cinema showing. In Spain, during the Civil War, newsreels sprang up supporting either of the warring sides: *España al Dia* for the Republicans, *Noticiario Español* for the Nationalists.

Newsreels had arrived at their most familiar form. Around ten minutes in length, a typical issue would contain some seven or eight stories, each with an opening title, overlaid by jaunty music and voiceover from an unseen commentator. The visual style was characterized by rapid editing and subservience to the commentary. Although the newsreels always privileged news that was worth *seeing*, they invariably foregrounded the word in how it was delivered. The tone veered between the portentous and the facetious, though the reputation that the newsreels acquired for flippancy and avoidance of hard news issues is unfair. The newsreels had to operate within an entertainment environment (the cinema) where they were not the main attraction. They could not afford to be challenging, because the exhibitor could always book one of their rivals instead. Economic reality, coupled with the inclinations of the newsreel owners, forced them into a passive state, ever eager to please and not to cause offence.

Nevertheless, the pusillanimity and frequent preference for the jocular of the newsreels in the face of some issues enraged many commentators, perhaps best expressed in Cecil Day Lewis's renowned poem 'Newsreel' about cinema audiences being anaesthetized by the newsreels against the realities of the Spanish Civil War: 'Enter the dreamhouse, brothers and sisters ...'. The exciting American news magazine (not strictly speaking a newsreel) *The March of Time*, with its bold, journalistic attitude towards issues of the day, was held up by many as showing where the newsreels were going wrong. But the newsreels knew that they were popular as they were, and saw little reason to change. They had the genuinely popular touch, and many dedicated newsreel cinemas sprang up in the 1930s for audiences who wanted to see newsreels and their like alone. The rise of newsreel cinemas also seemed to be fulfilling the newsreels' cherished dream, to be accorded the same respect as their rival news medium, newspapers.

The newsreels were often compared to the newspapers, not least by themselves, but a crucial difference was that the newspaper was a private choice, whereas the newsreel was part of a programme, and indeed seldom the chief reason why someone went to the cinema, unless it happened to be a newsreel cinema. Comparisons with newspapers are also awkward when it comes to considering the newsreels' impact. It has been argued that, for example, each bi-weekly issue of *Gaumont-British News* (Fig. 6.1) in 1936 reached 3.5 million people, comparable with the number of people who might read a popular British daily newspaper.[2] This is a useful rough gauge in determining impact, at least in terms of highlighting the newsreels' significance as a news media phenomenon, but it is ultimately misleading.

Figure 6.1 Gaumont-British News camera team, *c.*1936

A newsreel was not a newspaper; it delivered its information in a quite different way, it was not consumed in the same way. The two media were wholly different in the environment in which they were potentially apprehended: the one (newspapers) clearly more expansive in its use of words and in its range of subject matter, and a medium which did not constrict the consumer in terms of time spent; the other (newsreels) a time-delimited swift report on events, with an emphasis on headlines, easy summaries and of course the visual in motion. One has to say 'potentially apprehended', because how a newspaper *might* be read is not necessarily the same as how it *will* be read. We skim over headlines, we read the beginning and ending of reports, we ignore those stories that do not interest us, we look at the pictures, we create our own news construction out of the multiplicity of choices offered by the newspaper form.

The newsreel was not, on the face of it, as complex a medium as the newspaper, and effectively it could be consumed only in the one way. But the newsreel did not exist in isolation. It was consciously constructed as part of a chain of news provision, serving the needs of a cinema audience already informed about what other media, namely newspapers and radio, had determined the news should be. The newsreel can be considered in isolation, as any media phenomenon might be, but it is misguided to assess its impact as a medium on this basis. The newsreel was only one part of a wider process of apprehension of the news, of how the news of the moment was visualized and comprehended by a public that was being offered the news through a multiplicity of outlets, in a variety of forms, which then can be seen as contributing collectively to a wider form of the news.

Consider how one reads the news today. One may hear the radio news in the morning, or catch breakfast television. A journey into work might mean a newspaper, to be read in a variety of ways according to time available or personal interest. One can, of course, read a newspaper and consciously avoid anything that might be described as 'news'. Gradually one pulls together a composite picture of what one wants the news to be, seeking out confirmation through headlines, web pages, RSS web feeds, mobile phones and podcasts, according to one's degree of media literacy. We make the news what we want it to be. The point must be this has always been so – at least for the twentieth century and beyond. Of course, care must be taken in comparing the multimedia environment and its consumers of today with audiences in the age of the newsreels, but the choice of media was nevertheless there, and the newsreels were a product of a competitive and reflexive news media market.

Economic realities and the practicalities of film production and distribution determined what the newsreels became. They then adapted and flourished in that soil in which they found themselves planted. Newsreels knew that they were destined to be 'late' with the news, but they deliberately positioned themselves in a chain of news provision at whose head stood – at that time – the newspapers. The audience in the cinema already knew what the news was when they entered the cinema; they now wanted to see the pictures, to have brought to them a visualization of topicality. The newsreels completed the news picture determined by the newspapers, and as they grew in sophistication, provided a summary of events that was succinct or simplistic, according to taste. They did not act alone, and never saw themselves as acting alone. They were a link in a chain, an interdependent communication network that grew in complexity as illustrated journals, then radio, and in the post-war period television, added to the rich nexus of news media from which the public selected and determined their understanding of what was 'news'. And the newsreels were deeply conscious of this wider world of news provision into which they fitted, and conducted themselves accordingly.

The Second World War was the newsreels' hour. The cinema was, along with radio, the home for airing to a whole country the issues of the day. In the United States, the five main newsreels (*Fox Movietone, MGM, Paramount, Universal* and *Warner-Pathé*) continued as they were, independent of government, but each contributed content to a government newsreel, *United Newsreel* (1942–5), for overseas exhibition. In Britain, the government did not take over a newsreel as in the First World War, but the five British reels (*Gaumont-British News, Pathé Gazette, British Paramount News, British Movietone News* and *Universal News*) each co-operated closely with, and were censored by, the Ministry of Information. All contributed to *British News*, designed for overseas exhibition, while specialist reels such as *Worker and Warfront* (1942–6) and *The Gen* (1943–5) were produced by official sources for factory workers and the armed services. Elsewhere, centralized production was the norm. In Germany, all independent newsreel production was halted, and the existing services directed through a single newsreel, *Deutsche Wochenschau* (1940–5). In Spain the one official newsreel was produced, *NO-DO*, which had to be screened at all cinemas. It was founded in 1942 and lasted all the way to 1981 (losing its obligatory status only in 1975).[3] The *Luce* newsreel continued to be obligatory in Italy, and Japan followed the German model in subsuming all newsreel production under the one national service, *Nihon Nyusu*. There was no single newsreel service in the Soviet Union, but all production was nevertheless centralized. Towards the end of the war, and for some years thereafter, specialist newsreels were created for political re-education. The Anglo-American newsreel *Welt im Film* (1945–52) had a specific deNazification remit and was shown in Germany and Austria, while *The Free World*, produced by the Allied Information Service (the combined forces of the British Ministry of Information and the American Office of War information), was exhibited extensively (and compulsorily) in several liberated countries until the end of the war.[4]

The Second World War had seen the newsreels reach a peak of influence, by their sheer ubiquity and the power of the images they could convey. However, such power also helped to weaken them. Cinema audiences were exposed to news on a worldwide scale, and to issues and a diversity of cultures which changed their sense of what news could be. They knew more and expected to see more. The newsreels were unable to build upon this challenge. Firmly a part of the cinema programme, wedded to stylistic formulae which did not encourage them to embrace change, and most fundamentally tied to one or two releases per week, the newsreels found that in the post-war world they were losing the command that they held upon audiences.

A new medium arose to fill the audience need, television. Originally television news services were no more than newsreels themselves. In the 1930s the BBC showed *Movietone* newsreels rather than provide its own news broadcasts. In 1948 it introduced *Television Newsreel*, essentially its own newsreel (without newscaster), broadcast two times a week. It seemed that the newsreel had established not only the form of visual news, but also its ideal frequency. However, *Television Newsreel* moved to three editions a week by the end of 1950, and five editions a week by mid-1952. With the arrival of commercial television in Britain in September 1955, a new kind of newsfilm programme had emerged, with disparate stories arranged according to some kind of scale of importance, linked by a visible newscaster, and stripped of any musical background. The newsreels were now antiquated. They made various attempts to evolve stylistically, including talking heads, interviewers and vox pops, but they were inexorably tied to the distribution patterns of the cinema circuits, late with a news that the mass audience now expected to see on a daily basis.

Newsreels declined at different rates across different territories. In Britain, *British Paramount News*, *Universal News* and *Gaumont-British News* had all folded by the end of the 1950s, the latter turning into a magazine film series, *Look at Life*. *Pathé News* continued to 1970, and *British Movietone News* to 1979 (Fig. 6.2), sustained that little bit longer by government contracts. The last US newsreel, *Universal News*, closed its doors in 1967, and the final newsreel was shown in France in 1980. The *Movietone* and *Cinesound* newsreels in Australia merged in 1970, but finally ended in 1975. Elsewhere, the form persisted where there was public funding of some kind: *NO-DO* in Spain lasted until 1981; Belgium's *Belgavox* until 1994. In Japan, major newsreels were still being shown into the 1990s.

Figure 6.2 British Movietone News cameraman Norman Fisher

However, as the newsreels faded, so a new academic interest arose in their content (which had not existed while they were active). This came about through the prominence accorded to historical television series such as the BBC's *The Great War* (1964) and ITV's *The World at War* (1973), which used substantial amounts of newsreel footage, and which begged all manner of questions concerning the provenance and historical meaning of the newsreels. Nicholas Pronay, perhaps most prominent among the academics who engaged with the newsreels as a tool for historians, wrote two key essays on the newsreels in the 1930s for the journal *History*, which remain valid for the understanding of newsreels in an academic context, and for assessing the meaning of newsreels as a phenomenon beyond their allotted time as a fixture in the cinema exhibition space. Pronay writes:

> They are important historical evidence which deserves study. Not as records of events, but as records of what a very large, socially important and relatively little documented section of the public saw and heard, regularly from childhood to middle age. They are also primary evidence for the history of those wider developments which are brought about by the application of modern technology to communications and which have been once described by Professor Asa Briggs as: 'the changes in the ways of seeing and feeling and forms of perception and consciousness'.[5]

Pronay wrote of the newsreels as a phenomenon which had moved from the cinema to the footage library, and from news provision to the provision of historical evidence – not in the same way as textual sources, but as evidence of those issues to which the public were, or were not, exposed. This in itself is further indication of the protean nature of any example of the media in the modern communications landscape. Cinema news becomes history programming, with a whole new set of connections within the present-day visualization of history and its themes.

As Pronay implies, the newsreels indeed did not so much fade away and die as evolve into other forms, namely news agencies and footage libraries. The history of the British Commonwealth International Newsfilm Agency (BCINA) illustrates the pattern. Newsfilm agencies appeared in the 1950s to cater for the burgeoning television market. BCINA was formed in 1957, a trust formed by the BBC, the Rank Organisation, the Canadian Broadcasting Corporation and the Australian Broadcasting Commission. This brought together the Rank newsreel libraries of Gaumont and Universal, as well as *British Paramount News*, whose laboratories Rank owned.[6] BCINA marketed itself as Visnews, which became a major worldwide newsfilm provider, particularly to the BBC, for many years, until it was absorbed by Reuters Television, which in turn is now managed by ITN Source (previously ITN Archive). ITN, the leading British commercial television newsfilm provider since 1955, now finds itself custodian of most of the British newsreel heritage, holding the Gaumont, Paramount and Universal newsreels from its Visnews/Reuters inheritance, as well as the Pathé newsreel library, since it took over the management of the British Pathe collection in 2003. Today many of the major newsreel collections enjoy a second life in footage libraries, though the form in which they survive varies. Some are held as complete issues, some as individual stories; many mix unreleased with released stories, so it is difficult for the researcher to know whether or in what form the film was exposed to an audience; and many have no soundtracks, as only the images were deemed of value for resale – ironically, in view of the dominance of music and commentary for the newsreels in their exhibition form.[7] What the academic, or footage researcher, may uncover in a newsreel library may be at more of a remove from the experience of the original audience than simply through time.

However, academic investigation into the newsreels has dwindled since the enthusiasm generated by Nicholas Pronay, Tony Aldgate, D.J. Wenden, Paul Smith and others in the 1970s and 1980s, despite a wealth of newsreel footage having become readily accessible to academic researchers in recent years, with the release of the entire British Pathe library online in 2002, followed by the merged Gaumont and Pathé libraries in France (www.gaumontpathearchives.com) and comparable collections for German (www.wochenschau-archiv.de), Italian (www.archivioluce.com) and British newsreels (www.nfo.ac.uk and www.movietone.com). Newsreels are common on the web, but have a shrinking visibility on television, and this seems to have diminished academic interest in the sorts of evidence that they may provide. Producers of history programmes on television have turned away from the newsreels, believing that audiences no longer want or indeed empathize with monochrome, no matter how much it may denote a feeling of being actually 'there'. Instead, the blandishments of home movies in colour, dramatic reconstructions and digital effects are employed to persuade an audience of historical reality. For both academics and programme makers, it seems that meaning is tied to contemporary taste. There has been an increase in citations of newsreel coverage among those studying the media coverage of historical events, so at least there is a recognition among some that newspaper coverage alone is an inadequate gauge of public perception, but too often the newsreel is damned for its perceived inadequacies as concrete evidence, without due understanding of its integral form and function.

Recently, the American academics Kevin G. Barnhurst and John Nerone have provided a refreshingly different perspective on newspaper history, by looking beyond the study of its content (the traditional interest of scholars) to examine its particular form. Their book, *The Form of News*, looks at the ways newspapers present themselves – how they are written, how they are organized, how space is managed, the positioning of pictorial material, their typography, and offers a new analysis of the relationship between changing newspaper form and the function that the medium plays in society.[8] *The Form of News* suggests ways not only to examine newspapers, but also to consider other forms of news provision, including the newsreels.

According to Barnhurt and Nerone, the newspaper form holds a special sanctity which derives from its relationship to civic culture. Historically, the newspaper acquired, and has maintained, a particular respect as an instrument of democracy; a sacred mission often at odds with the newspapers' commercial operations. The newspaper, therefore, brings with it a set of assumptions and expectations. These are expressed in the form in which the newspaper presents itself, not simply in how it looks or arranges itself, but in how it represents itself to its public, and the ways in which it makes itself available.

The newsreel had no such roots in civic expectations. Its form derived from the demands of the cinema audience, which came to favour a mixed programme of film content spread over part of an evening. It was accepted, rather than approached with expectations demanded of it. Cinemas did not inevitably have to have newsreels: there was no positive demand for them as such, but they were popular, and they contributed to the variety of the programme. They were a part of an evening's entertainment, and being a small part meant they had little economic weight. There were recognized as part of the cinema programme, but they were not the reason why anyone went to the cinema. The expectations that they carried with them, that made them recognizable, were therefore quite different from the newspapers. They were expected to reflect an image of the news, to do so in an entertaining manner, and to do so at speed.

Speed is the newsreels' dominant metaphor. Stories about newsreel camera operators obtaining stories, amid much derring-do, are dominated by notions of speed – rushing the film to the labs ahead of your rivals, getting the hastily processed film to the major cinemas in a matter of hours, exhibiting films of cup finals on the evening that they took place, and so forth. Speed was in the length of the newsreels: one reel of film, ten minutes to display half a week's news, and not a minute to be allowed over that time. Speed is also in the presentation of the news itself, for to watch the newsreels is to be amazed at how rapidly their subjects pass by. Shots are seldom held beyond five seconds; the action is propelled along by urgent music and an insistent commentary. *Gaumont-British News* included a feature, *Roving Camera Reports*, which featured almost subliminal snapshots of action from around the world. The viewer today must wonder at just what the audience of the time was expected to make of such a rapid flick through the stories of the hour. Scholars citing the newsreels in media histories sometimes express puzzlement at how fleeting and seemingly uninformative newsreel stories of matters of great moment can appear to be. The *Gaumont-British News* report on the bombing of Guernica lasts for some 30 seconds – four or five shots, then on to the next story, a *Roving Camera Report* on relay races in Philadelphia (*Gaumont-British News* no. 350, released 6 May 1937).

The speed of the newsreels naturally ran contrary to a deep consideration of the news. Instead, they offered a summary or a checklist of news events, and crucially a visualization of those events. They conferred visibility upon things, even while whisking away the one image to be replaced by the next. Their part in that chain of news provision was to confirm visibility and thereby to confirm newsworthiness, for to be seen was indeed to be in the news.

Newsreel form meant a rapid succession of the familiar. It meant a triumphalist opening with a logo akin to those found at the start of feature films, then a series of stories each introduced by a headline with music appropriate to the tone of the story, and then the off-screen commentator reading out the story's salient details and import, the words tightly edited to the whirl of images that collectively made up each story. There was no particular order to the stories – newsreels were often constructed simply in the order in which the individual filmed stories came out of the labs – only a shared topicality to bind them together, before another triumphal sign off, and then lights up and time for ice creams. Its form supplied meaning, because newsfilm of itself is otherwise meaningless. It needs to have a focus, to have a specific relevance to an audience situated in a particular time and place for it to take on news relevance. It has to be recognized as news.

In being part of the nexus of news provision, the newsreel was not so much at the end of the chain as but one link in a circle; that is, to be in the newsreels was to be news, and such visibility would then be reflected in other news media. People became famous through being shown on the newsreels. Newspaper photo pages emulated the multiplicity of shots and angles offered by newsreel coverage. Newsreels contributed to what was seen, how it was seen, and to a sense of news in motion. Their very existence contributed to a transference and accretion of meaning which one can now read in the distribution of digital news images today.[9]

To see how the news moves today across computer networks, and in particular how it presents itself on the web, is to see what a complex and elusive concept news really is. No one form can hold it, and web news pages do not really present the news as such, but instead offer an infinite variety of news options through an extensive system of cross-references. The news is to be selected from across the choices made available on the web page, or else is at a remove, always a hyperlink away. Web pages offer the news in abundance, inviting selection, expecting one's personal interpretation, so that for every person a news web page is different, just as the BBC News web page promises to be updated every minute of every day. Taking things to their logical extension, one can never get the same news twice.

This sense of news in flux, of complex interaction of the visual, the textual and the aural, of the necessity of choice, not only is there for all to see in the present digital environment, but also offers a useful means to consider the provision of news across the past century. The news has always been like this, ever since technological revolutions at the end of the nineteenth century gave us those modern communication systems whose news products interact to give us our composite picture of the news. The news is, ultimately, a personal choice made by the consumer who selects and aggregates their particular news agenda from the variety of communications media on offer. There is a larger news form, outside any one medium. What exactly that form might take is hard to describe or to summarize, but a characteristic point must be the need for the news media to fit in with the human daily round, with certain regular kinds of human activity – like switching on the radio in the morning, like reading a newspaper on the train, like going to the cinema and seeing the newsreel. All contribute to a particular news environment. None can be seen in isolation.

The newsreels were a popular and pervasive medium, viewed by millions, successful for a number of decades in one public form, and then successfully remarketed in other forms which traded on their value as news footage and the stuff of historical documentaries. They are an inescapable part of our visual past, because no one is going to be able to go back in time and film those stories again. History programmes may currently prefer colour film of the past, or even the colourization of monochrome; they may for now have a taste for recreations with actors to the hackneyed newsreel footage of old; but the newsreels are still the benchmark. The much-hyped television programme *Virtual History – The Secret Plot to Kill Hitler* (Discovery Channel, 2004) which boasted of its ground-breaking combination of digital trickery and actors to recreate the unfilmed Hitler bomb plot, nevertheless felt the need to present the key scene as if filmed, with self-conscious handheld shots, as though willing the cameras to have been there. Faux newsreel footage, with added-on scratches, is a new cliché for the times. Film is still the measure for how we expect the past to be visualized.

The newsreels inform our picture of the past so effectively because they were central to the creation of that visibility in the first place. They reflected the news, but equally they made it, and they cannot be ignored nor can they were viewed in isolation. They were an integral part of the bigger picture; indeed, it could not be otherwise, since the news must always be greater than those individual media that play their part in carrying it. It is necessary to look at the newsreels again: to consider their form and function, the ideal that they represented against the reality of any one individual reel. The experience of viewing newsreels should be used to gauge other news media, not just looking at the content but at the mode of delivery and the complexity of its meanings. The newsreels are an indivisible part of the visualization and comprehension of the news agenda of the twentieth century.

Notes

1   Luke McKernan, *Topical Budget: The Great British News Film* (London: British Film Institute, 1992), 64–7.

2   Nicholas Hiley, 'Audiences in the Newsreel Period', in *'The Story of the Century': An International Newsfilm Conference – Papers, Presentations and Proceedings*, ed. Clyde Jeavons, Jane Mercer and Daniela Kirchner (London: British Universities Film and Video Council, 1998), 60–1.

3   Rafael de España, 'Newsreel Series: Spain/Portugal', *Encyclopedia of Documentary Film,* ed. Ian Aitken (New York: Routledge, 2006), 988.

4   Roel Vande Winkel, 'Newsreel Series: World Overview', in *Encyclopedia of Documentary Film*, 988.

5   Nicholas Pronay, 'British Newsreels in the 1930s: Their Policies and Impact', *History* 57, no. 189 (1972), 72.

6   James Ballantyne, ed., *Researcher's Guide to British Newsreels*, vol. 3 (London: British Universities Film and Video Council, 1993), 19–20.

7   For example, the *British Paramount News* and *Universal News* libraries within ITN Source mostly lack soundtracks.

8   Kevin G. Barnhurst and John Nerone, *The Form of News: A History* (New York: Guilford, 2001).

9   See Ron Burnett, *How Images Think* (Cambridge, MA: MIT Press, 2004).

7 Documentaries: a gold mine historians should begin to exploit

by Pierre Sorlin

Documentaries were long despised by film-buffs, who considered them a patchy aggregate of hastily shot pictures screened at the beginning of the film show, while latecomers were noisily looking for their seats. Few realized that, every year, ministries, industrial companies, railways, armies, hospitals, tourist offices, museums and scientific societies shot or sponsored hundreds of informative shorts. After decades of contempt, factual films triggered interest and curiosity in the last third of the twentieth century; there were specialized film archives to collect them, festivals to diffuse them, university chairs and books to study them.[1] Supreme recognition: in 1995 Hollywood instituted an 'Oscar' for documentaries.

Such reappraisal was, in itself, a small historical event and, as such, it deserves a few words of explanation. In the 1960s the new lightweight cameras allowed militants struggling against capitalism and its ideology to circulate cheaply made shorts. After the Cuban revolution, young Latin American intellectuals used documentaries to call for social change. In their wake, US and European students filmed lampoons against the classical Hollywood cinema, which they regarded as plunging people in a false state of happiness. Then, during the Vietnam War, pictures served to counter the television news. The 1960s and 1970s were the heyday of political documentaries. In the 1980s, television channels proliferated; being obliged to broadcast day and night, they put in the aired films already available and ordered informative movies, which cost less than fictions.[2] Between 1990 and 2000, the number of hours given to documentaries on British channels increased fivefold. Meanwhile, public or private institutions, which had produced filmed adverts, realized that the valorization of their 'heritage' would be a costless form of corporate sponsorship. The accessibility of efficient digital cameras drove amateurs to film their surroundings, in the hope that television would buy their images. In our days the production of factual movies exceeds by far that of feature films.

What are documentaries?

The very plethora makes them difficult to define.[3] Things were simpler in the first half of the twentieth century. The difference between documentaries and news bulletins was and is still obvious: cinema and television news deal with what has just happened; their style is fast moving and simple; their images are scarce and often come from only one informant, be it special correspondent or agency, while documentary filmmakers stand back and have the benefit of insight. It is less easy to separate fictions and documentaries.

The reference to 'reality' is of no avail, for few, even in the early years of cinema, thought that films mirror the outer world since the framing of an object or a person is already a manipulation. At best, a documentary is a 'processing' of reality. Length is not a criterion since there are 20-minute fictions and two-hour documentaries. Fictions have been acted out, but such is the case of many factual movies. Sometimes it is necessary to represent what could not be filmed. In 1945, for example, Australian dock workers blocked the ships carrying Dutch soldiers to put down the Indonesian insurrection, but no camera was present. With the help of unionist members, Joris Ivens' *Indonesia Calling* reconstructed the episode 'as it had occurred'. In other cases a short fiction spares long explanations. Actors playing husband and wife told, in Robert J. Flaherty's *The Land* (1942), what the problems of irrigation were, and how they could be solved. Their dialogue was less boring than the speech of an anonymous voiceover. The dividing line is often very thin; it would not be arbitrary to label Flaherty's *Louisiana Story* (1948) 'fictional documentary' (Fig. 7.1) and John Ford's *Grapes of Wrath* (1940) 'documentary fiction'. It is mostly the way of filming and the aim of the work which help distinguish traditional documentaries and fictions. The former focus on a topic, which they present and explain; their actors do not impersonate heroes, only anonymous people; and the films do not indulge in thrill or suspense.

Figure 7.1 *Louisiana Story*

Television has made the definition of documentaries much more problematic. How shall we characterize reportages which, concentrating on a question, a situation, a drama, offer genuine data, testimonies and opinions hurriedly gathered and shakily assembled? Are they protracted newsreels or clumsy documentaries? And what about 'reality shows' with their 'real' people, filmed in an artificial situation, which does not match their ordinary life, but tallies with their illusions? The only possible answer is an empirical one: a documentary is a film to which spectators attribute with authenticity and informative power; but such a fuzzy characterization shows how embarrassing the problem is. The image processing is generally a good guide, because operators working for 'reality' programmes have been told simply to take pictures, regardless of how unmeaningful they are, whereas documentary filmmakers try to select the best and most evocative shots. The context in which the film was made and diffused, the public at which it was aimed, the places where it was projected, the function for which it was destined are also an important clues. But sometimes the question turns out to be irrelevant because the television 'flow' ignores clear-cut divisions. Television surrounds us; it has its part in our daily activities; we pay it a fluctuating attention, and do not watch it as our grandparents watched films in the middle of the twentieth century. In the same way, people invited to appear on a television programme do not talk as their predecessors did when they were interviewed for a film. The meaning attributed to audiovisual productions has changed with the passing of years so that historians, when using them, must take into account the state of mind of those who shot and those who looked.

What kind of history on the screen?

How shall we cope with such a wide, ill-defined cluster of movies? All those interested in informative films face the same quandary and their answer depends upon the questions they raise according to their speciality and specific interests. Historians, for their part, focus on three main themes: what can they do with historical documentaries? What kind of information will they draw from mainstream documentaries? How will they insert documentaries in their works?

History has long been an excellent, inexhaustible topic for documentary filmmakers. In 1917 British Pathé inserted in their newsreels a *History of the Great War*, an assemblage of part of their own records. Inexpensive and quickly edited, the series met with a warm response. Spectators were keen on seeing again, with hindsight, what they had seen without knowing what was about to happen. It became technically easier to make compilations in 1926, when Eastman Kodak found a method of making duplicate negatives from film prints. The producing companies understood that retrospectives were a gold mine; compilations flourished and were distributed throughout the world. First came *The World War through German Spectacles*,[4] and *Verdun: Visions of History*.[5] These films are worth mentioning because they encompass all the shortcomings and approximations of most war films to come: they indiscriminately mix pictures from extremely different origins, compensate for the lack of documents with dramatizations and animated diagrams, mechanically follow a chronological order, and inflict on viewers an authoritarian and often simplistic commentary.

Television made it possible to edit extensive series, broadcast through an entire season. Agencies like Visnews bought the libraries of newsreel companies, which had ceased production in the early 1960s, and constituted vast archives to service history filmmakers. Hundreds of history programmes overwhelmed television screens. Naming them all would be tedious, but let us remember a few big hits: *The Great War*,[6] *World at*

War on the Second World War conflict,[7] *People's Century*,[8] an original calling-up of the main events and most significant changes that marked the twentieth century. With the passing of time some mistakes have been corrected but *World at War* and *People's Century* are still chronologically ordered and there is no dominant point of view. Testimonies alternate with original pictures and provide the opinion of great figures as well as of ordinary citizens. Yet the filmmakers did not succeed in introducing anything not based on images. In both series the economic crisis of 1929 and its aftermaths was illustrated but not explained; Hitler's rise to power, the territorial ambitions of Germany and Japan, the latent conflict between Japan and the United States, the internal situation of the 'Soviet bloc' did not appear clearly.

Despite their positive qualities, these programmes point up the limitations of historical documentaries: films show, they put their spectators in front of a reproduction or an imitation of some events, but they can neither comment nor enlighten them. Images are unable to tell things as simple as: 'this happened in 19__' or 'European powers were competing for the domination of Africa', let alone abstract notions like capitalism, imperialism or socialism. Attempts at presenting visually abstract concepts turn out to look ridiculous. *World at War* alternated shots of the persecution of the Jews in pre-war Germany with home movies made by Eva Braun. Such elementary opposition between happiness and misery misses the main question: why did ordinary Germans blame a minority for their difficulties and harass it? Filmmakers who want to convey such ideas have recourse to words; thus most 'history films' are traditional, and more or less accurate history lectures, completed by pictures.

Oral information in historical documentaries is poor and of little avail for specialists who, on the other hand, may learn a lot by looking at the pictures, especially when researchers have dug them out of unclassified archives. For that reason historians must become acquainted with the analytical study of visual sources. Filmmakers make use of three types of material: archival footage, images taken during the shooting of the film, and re-enactments. A few common-sense rules will help to detect dubious footage. Films shot in actual circumstances – during an accident, a race, a public demonstration, a riot, a battle – are likely to be badly lit, badly exposed, shaky and possibly out of focus. If a cameraman filmed with his back to the enemy, as his own side was advancing, how did he get there and survive? If a movie shows a gun firing and then the explosion of the shell, how was it possible to film the blast behind the enemy's front line? Obviously the second shot does not match the first. In 1931 a German director decided to reconstitute the taking of the Douaumont fortress, conquered by the Germans in 1916.[9] Soldiers who had taken part in the attack agreed to participate. Weapons, uniforms, locations – everything was authentic – but the forgery is visible at first sight. No operator, following the attackers, would have been able to take such perfect pictures, well framed and well lit. Moreover, there is no sign of fear on the part of the men, who were not afraid of a simulated resistance and know exactly what they had to do. In other cases a close examination of the street scenes for the types of car and clothing styles, of the posters, adverts, street signs, that is to say a translation to images of the critical procedures used for written documents, allows one to separate faked pictures from actuality footage. It must also be borne in mind that operators may have been hindered by external pressures. Some pictures are 'genuine' but have been shot under the control of an authority: politicians who take great care over their appearance, the owner of a factory who does want to show how timeworn the plant is, a military censor afraid of leaking information to the enemy.

Even authentic documents may turn out to be not valid. *Night and Fog*,[10] released in 1955, was the first documentary that attempted to provide a synthetic view of deportation and of the Nazi death camps (Fig. 7.2). Its impact was considerable and it is still frequently broadcast to keep alive the memory of the camps. The film begins with a sequence showing people wearing the yellow star, carrying luggage, going up into wagons, the doors of which are closed by German soldiers. The film does not tell spectators that these pictures, the only available ones about a deportee train, had been taken by order of a German officer who wanted to show that everything was under control. There is neither hurry nor violence, whereas all survivors say that such transports were extremely brutal. After the closing of the doors, another shot shows the arrival of a deportation train, in front of German soldiers. This picture was borrowed from *Ostatni etap* (*The Last Stage*), a Polish fiction film about death camps shot in 1946. A few minutes later we see people gathered in a big stadium; the commentary explains that these are Jews arrested by the French police in July 1942 and destined to deportation. Many details prove that the filmmaker was mistaken; these individuals were actually collaborationists arrested after the liberation of France. Thus, out of three images likely to move viewers, one is a re-enactment, one a propaganda shot, and one a real document erroneously labelled.

Figure 7.2 *Night and Fog*

Film directors are generally more sensitive to the quality of the pictures, their clarity and their expressive power, than to their accuracy. In the final credits they sometimes tell where their documents come from, but this is not enough. Historians require more precision and should ask for the insertion, in DVDs, of virtual notes, accessible only on request, indicating the date of every shot and the place where it can be viewed. That would not be difficult, but producers will not bother, unless they are insistently solicited to do it.

Crude facts and inventiveness

History films are a mere island in the ocean of informative movies. Historians wonder whether documentaries might come in handy for them, but often they do not know how they could take their bearings amid such a huge quantity of pictures and may doubt that moving images would facilitate their research. Catalogues are available on the Internet, but a title, a summary and a few clips are not enough. The lists, however precise they are, convey neither the continuity of the motions, nor the effect produced by the editing process. It is true that throwing oneself into a register is hopeless; what researchers should do, before looking for audiovisual sources, is think over their objectives and ponder the kind of data they are looking for. Films may intervene at different stages of the investigation, and provide various forms of wisdom about a theme. In this chapter I shall consider three types of documentaries: factual, critical and propagandist. Such a division sounds extremely schematic, for there is no clear-cut division between these series; some become prevalent in certain circumstances (propaganda films during wars) and others vanish for some time. All I hope to show is that a lucid definition of one's aim makes it easier to select the convenient cinematic documents.

Factual films, 'films of facts', describe objects or, more often, practices. A very straightforward example, *Gjutarna, the Iron Founders*, a Swedish film,[11] synthesizes a work day in a small, artisan foundry. The workers have been asked to act out their jobs as usual. The film takes us through every single step of the process; there is no commentary or musical score other than the few words exchanged amid the men and the noise of the tools. Since the beginning of the twentieth century hundreds of companies, public (such as the famous state-sponsored offices, EMB Film Unit, GPO Film Unit, Crown Film Unit, from 1927 to 1952) or independent (the US Workers' Film and Photo League and Frontier Films), have shot thousands of movies about labour, traffic and trades, providing us with images of activities which, quite often, had not changed for decades, but which no longer survive. There exists a prodigious library of moving pictures that allow historians to 'observe' directly practices which interest them, instead of painfully trying to reconstruct them from written documents. Such pictures record not only the methods, but also the relationships between the workers, their tools and their industrial tasks. It is extremely important, now that automation has reduced the number of employees and limited the nuisances in most workshops, to look at the experience of those who worked in factories six or ten decades ago, and to realize how tiresome and dangerous their job was. By simply watching *A Day in the Life of a Coal Miner*,[12] we understand, much better than by reading any literary account, the physical environment and the activities of a mining community before the First World War. By comparing it to *Coal Face*,[13] we see what changed during the 25 years that separated these pictures, and what had remained unaffected. Then, *Living on the Edge* takes us to the end,[14] to the extinction of the mining community after the 1984 crisis; three films point up an entire century.

Beside industry we could mention agriculture, commerce, urbanism, leisure, school or prisons. As a matter of fact, few fields have escaped the eye of filmmakers. Documentaries bear witness to people doing things; they show paralinguistic forms of expression, gestures, demeanour and dresses, and give an idea of the various systems of communication which convey signification for the members of a neighbourhood. By illustrating the exchange of civilities, the forms of deference and scorn, they help to identify basic social relationships and expose the meaning that participants attach to them. To take but one example, the differentiation of sexes and the preponderance of the male point of view appear explicitly in *A Complaint of Rape*,[15] where a young lady who tries to lodge a complaint in a police station comes up against the ironic scepticism of the constables. Such pictures do not show 'historical events'. Nevertheless they are expressive manifestations of social interaction and should interest researchers because they unveil the background of interpersonal relationships. Thanks to documentaries, historians and sociologists could 'visualize' the functioning of world societies in the twentieth and twenty-first centuries.

That is to say that we have at hand a rigorous representation, and nothing more than that. Documentaries do not offer 'slices of life', but merely visual accounts of parts of the world. Even a movie as honest and simple as *Gjutarna, the Iron Founders* is biased in many respects; the course of action has been reduced to the most significant gestures, and the desire to make everything easily comprehensible has led the director to synthesize and eliminate many stages of the whole process. Distortion is unavoidable. Many factual films, sponsored by public authorities or major firms, were aimed at officials or at the staff of the firms, executives, workers, agents or marketing people, and gave them practical instructions. Consider *Health of the Nation*,[16] co-produced by the Pearl Assurance Company and the Ministry of Health, in the hope that a play for hygiene, maternity care and a wholesome diet would reduce health expenses. It was an education tool, meant to teach people how to behave soundly, and it showed what ought to be, not what happened really. More than mere advertisement, many of these films were communication tools projected in fairs or exhibitions, mainly for the prestige they brought to an enterprise or a ministerial office. Citroen, the French car manufacturer, provided the funds for the crossing of Africa, then of Asia, by people who filmed their trips. Such pictures promoted both the generous company and its vehicles. Some big corporations paid for movies extraneous to their activities but likely to popularize their trademark. LNW Railway provided the money for *A Day in the Life of a Coal Miner*; British Gas and Coke Company financed movies about lodging; Esso subsidized films on agriculture, irrigation and land reform. Consciously or not, the sponsors tended to paint a black picture of the situation they wanted to modify and to overestimate the results of their own intervention.

Those who shot the films were anxious to please their clients and eager to seduce the public. Some thought that spectators would be bored by crude facts and found it necessary to embroider them with a musical score and 'aesthetic' pictures. Historians will find in *Drifters*,[17] a film about herring fishing out at sea, information about an activity which changed drastically during the second half of the twentieth century (Fig. 7.3). They will observe the boats, the fishing techniques, and the way herrings were processed but they must be careful, for the filmmakers were intent on embellishing their movie and did not hesitate to distort their presentation. Some shots, chosen for their poetic value – waves rolling over the beach, gulls flying over the horizon, men sleeping in an

Figure 7.3 *Drifters*

expressionist light – mitigate the harshness of the work. More seriously, the fishermen who appear in the movie were selected because they looked handsome and were not representative of their comrades; in addition, the processing of the fish was not filmed on board, as it should have been done, but in a studio. Such a film is a free interpretation, not a genuine testimony, and this case is not exceptional; artistic claims and the desire to display good, clean pictures have altered the authenticity of many 'factual' films.

It is all the more tempting to 'embellish' when the film is shot in faraway countries, not likely to be familiar to the viewers. Given that spectators are not in a position to raise objections, ethnographic movies, movies dealing with human communities living outside the capitalist, industrial sphere of production, tend to intermix genuine takes with fancy ones. *Nanook of the North*,[18] the first systematic visual enquiry about a 'primitive' group, met with a tremendous success and remains a symbol of visual ethnography (Fig. 7.4). Focused on a family of the Itivimuit Eskimo tribe, chiefly on the father, Nanook, the film did not present the point of view of the natives, but the approach of an American impressed by what he fancied the proximity of the Eskimos with a yet unmodified, intact nature, and their constant fight against a ruthless environment. As a matter of fact, his picture described more a relic of an archaic, timeless civilization than the actual life of an ethnic group in the polar circle. To make his story more gripping and convincing, the filmmaker did not hesitate to fake. He gathered around Nanook people who were not his relatives but were of good appearance, and he made the man catch a walrus with a harpoon, although Eskimos had long adopted rifles.

Figure 7.4 *Nanook of the North*

When filmmakers work in a foreign country, where they have to rely on local informants, who is fooling whom? In this case, was it the American who made Nanook play the part of a primitive, or Nanook, who cheated the filmmaker by means of methods that were not his, but his father's? Shooting always implies a bargain between enquirer and testifier. Brazilian director Eduardo Coutinho wanted to interview a young girl for *Babilônia 2000*, a documentary about the *favelas* of Rio de Janeiro. When he arrived the girl was combing her hair. 'Don't,' he said, 'it isn't necessary.' 'I see,' replied the girl, 'you want to film poverty live.' He had his (preconceived) idea and she had hers; the film was a compromise.

Does it make sense to exploit documents in which the strategies of the witnesses interfere with those of the reporter? It all depends upon the historian's purpose. *Nanook of the North* does not document how Eskimos lived when the picture was shot; the character mimics what he had seen when he was young. But if he had been interviewed about his father's practices he would have mentioned the harpoon and his verbal depiction would be less telling than the pictures. Because the steel harpoon was quite recent, Nanook's testimony is valid for only a short period, roughly the second half of the nineteenth century, not for the totality of Eskimo culture. Ethnographic films are snapshots taken in specific circumstances, and, correctly questioned, they may be extremely helpful.

Communicative emotion

In presenting *Ghosts*, a film about the dramatic situation of Chinese illegal immigrants to Britain shot in 2006, scriptwriter Jez Lewis revealed that he had intended to be 'highly offensive'. What differentiates militant pictures from factual ones is that the former are aimed at proving something. They are not necessarily aggressive, but their objective is to persuade. But whom, and of what? There are, I think, two main categories: censorious reports, which we shall consider first, and propaganda films.

Critical pictures transcribe an emotion. Someone, having been shocked or impressed by an event, or a drama, such as the difficult condition of a community, expresses her or his feelings by means of images that speak for something and against something else, in order not only to inform but also to provoke a reaction. However, what ought to be done is not pointed out; viewers must interpret a message communicated by plot, images and editing, or by irony and dramatic contrast. For instance, *The VW Complex*, a German film about the Volkswagen manufacturing factory,[19] begins with the making of a car, in the context of contemporary economic activities, and progressively introduces an opposition between the comforts of modern life, symbolized by motorcars, and the human cost of industrial production.

When images are lacking, indictment films tend to become more lampoons than movies. In the absence of documents *The Thin Blue Line*,[20] an attack against police and justice system for a overly hasty enquiry and the condemnation to death of an innocent man, illustrated with stereotyped pictures the 'building' of criminality, and re-enacted some scenes according to the contradictory statements of the witnesses. But, generally, the visual material is genuine; historians can exploit it, provided they take care of the distortions introduced by the editing process and the commentary. The destitution and backwardness of some districts of rural Spain denounced in *Las Hurdes: Land without Bread* were unquestionable;[21] the pictures of puny, unhealthy people, wretched living conditions, miserable crop and dangerous roads were not faked. Yet, in order to reinforce the message and strongly impress the public, the film had recourse to appalling views, authentic but not necessarily representative, such as the hand beheading of roosters, or wasps devouring the corpse of a donkey. As it is, this movie cannot be projected as a description of Estremadura in the first half of the twentieth century, but the removal of a few shots and a less systematically dramatic linking of the sequences would make it a valuable document – again, one about a definite area at a precise epoch.

Cinematic denunciations were rare before the television era. Nobody wanted to distribute films like *Las Hurdes*, so militant filmmakers preferred to work for political parties, which would circulate their films among their members. Television networks changed that. They needed programmes likely to attract spectators, and reports about ongoing concerns could meet with a warm response. The BBC broke new ground when it launched, as early as 1952, *Special Enquiry*, a series of documentaries dealing with present-day issues, and most channels followed its example. By focusing on contemporary problems and interviewing 'ordinary citizens', television investigation, willingly or not, adopted a critical stance. When a man who had accepted the presence of a camera spontaneously told a friend: 'It seems to me that it's all right for blacks to kill blacks. That's perfectly okay. They can just kill as many as they want and nobody's going to say a word about it,'[22] this was not 'information', but a rather crude look at a sector of British society. The mere fact of letting people express themselves on the screen, instead of printing their words in papers, modified radically the coverage of events. A documentary such as *Death on the Rock*,[23] a report about the killing of three Irish Republican Army (IRA) members by the police, had a strong impact because witnesses demolished the official version of the facts, theoretically built on direct testimonies. Video diaries, born from an emotional reaction, are now invading television channels and the Internet, lightweight cameras and mobile telephones give amateurs a chance to diffuse hurriedly edited images about dissidents in Byelorussia, the closing and dismantling of viable factories in Italy, violence against civilians in Haiti, a hijacking in Latvia, or clandestine immigrants from Africa coming ashore on European coasts. However poor they are, these pictures disclose disturbances and sufferings, which sounded far away and abstract when they were only reported in the press.

Enforcing conviction

Unlike critical films, propaganda documentaries have precise purposes. They must convince, if necessary by appeals to emotion, must maintain self-confidence and dedication at home, must gain the sympathy of neutrals, and destroy the enemy's morale. Theirs is a willingly distorted version of events. In 1939 the British Ministry of Information, in its instructions to filmmakers, stated that: 'Shots of our soldiers laughing or playing football must be cut out of all newsreels and documentaries sent to France.' Dictatorships are blamed for their systematic recourse to misinformation, but democracies do not shrink from half-truths, even while they may try to avoid blatant lies. In this field there is no limit; even allies must be supplied with cant images.

History books give much space to propaganda. It is partly because there are plenty of details about state institutions such as the US Office of War Information or the German Reichsfilmintendanz. It is also because describing a film about work in coal mines is problematical and the effort impoverishes what is alive and telling on the screen, whereas enumerating and criticizing watchwords is easy. Propaganda is a variety of advertisement; everywhere, under liberal as well as under authoritarian regimes, it capitalizes on repetition, commonplaces and irony. Reiteration is fundamental; the meaning directly, insistently communicated to the public by a narrator, is often confirmed through the words of interviewees. A clever alternation of past travel accounts and contemporary annotations made *Song of Ceylon*,[24] poetic evocation of the big island, into a hymn to the beneficial effects of British colonization. In *The Principal Enemy*,[25] a Bolivian radical film, the foe, that is to say Yankee imperialism, is first named by the Indian narrator, then denounced by testimonies that explain how US-trained Bolivian militiamen kill rebellious Indians, a message underlined in a caption that closes the movie. Clear-cut good-and-evil stereotypes – the unrelenting contrast between affluent capitalists and miserable proletarians or between backward country people and avant-garde workers in Soviet pictures, the hook-nosed American bankers in Nazi movies, or the cruel, ruthless, Japanese in US war films[26] – all reinforce the message.

Money worries, patriotism and the quest for success led well-known filmmakers into propaganda work. A few documentaries are deemed masterpieces or at least important stages in the course of cinema history. *Triumph of the Will*, which showered praise upon Hitler on the occasion of a huge Nazi meeting, and *Olympia*,[27] an account of the Berlin Olympic Games, are often mentioned for their bright play on collective motions and their aesthetic enhancing of human body. Another Nazi film, *The Eternal Jew*,[28] is a paradigm of totally biased information; the Jews of Prague were obliged to play a degenerate people living in abject poverty. Among the countless Soviet films, the reconstitutions of the October Revolution and *The General Line*,[29] a fiction about the mechanization of agriculture, stand out.

What is the historical value of these works? They are not of paramount weight, but it is clear that the circumstances in which the films were shot and exploited may be of interest. For example, five weeks after the liquidation of his main rival, Ernst Röhm, Hitler commissioned a film showing that he was the undisputed leader of Nazism, and we cannot pass over it in silence but our interest is in the making of the movie and the internal conflicts inside the party, not the content of a film, which does not tell us anything useful about the Nazis, their objectives or their practices. Propaganda works are ideological artefacts; what they are meant to convey is more quickly found in written sources than in long sequences filled with predictable slogans and conventional images.

Did not such works exert an influence? The power of propaganda is a blind spot in historical research. As far as we know, an isolated item, a book, a speech or a film, has little if any effect on individuals. Rather, changes in opinion result from the conjunction of various factors. In 1940, how many of those in Germany who saw *The Eternal Jew* were already fanatic anti-Semites; how many were hesitant people; and how many opportunists who thought that two hours in a dark room would enhance their career? Similarly, in the heated years that followed May 1968, revolutionary documentaries shot in Latin America, such as the Argentine *The Hour of the Furnaces*,[30] circulated on university campuses along with protests against the Vietnam War; but did they persuade students who were not yet convinced? Sometimes the impact of a movie is plainly attested, as the case of the Bolivian *Blood of the Condor*,[31] which showed how, in a US Peace Corps maternity hospital, Indian women were being sterilized, while they thought they were cured. Dubbed in Indian languages of the Andean area, the film was projected in villages; members of the Collective organized public debates and obtained new testimonies about the Peace Corps's practices. The scandal became so manifest that the Peace Corps was obliged to leave Bolivia, but this was an extremely specific case, dealing not with political ideas, merely with a well-circumscribed operation. Against all expectations, propaganda films are not of much value for historians.

And now – what?

There are useful and useless documentaries. I hope I have helped to shed light on the issue, especially by warning against false friends, but that alone is not enough. I would like to conclude with more constructive propositions. I assume that historians, being used to studying the context in which texts were produced, will do the same for films. Once they know when, why, by whom their films were shot, how should they proceed?

A first step will consist in identifying the way the documentaries talk to their spectators, in order to catch their attention. We shall single out four types of addresses: explanatory, interactive, open and subjective. An explanatory film guides its public from start to finish. A presenter or, more often, an all-knowing voiceover, explains what can be seen on the screen. Take for instance an extremely popular, often broadcast television documentary *Victory at Sea*:[32] there are no sounds, other than a gripping musical score. Nor are there testimonies; everything is told by a narrator who describes with authority an inescapable march towards victory. In the interactive address, the introducer or the voiceover, instead of telling what is happening or giving their point of view, tries to provoke reactions on the part of filmed people by raising queries and signalling arguable points. In *Housing Problems*,[33] for instance, poorly lodged people are invited to make known their grievances, poorly lodged people describe the difficulties they must face. By contrast, the open address imposes: no explanations or commentaries. Spectators are meant to follow the actions, listen to the words uttered by the characters, and structure or put what they perceive in the right order. *Primary*, a documentary about the 1960 debate in the Democratic presidential primaries between Hubert Humphrey and John F. Kennedy,[34] presents aspects of the contest as they were filmed, without any additional remarks. Finally, the subjective approach films people acting and speaking freely, with no interference from the filmmaker. *Mendel Schainfeld's Second Journey to Germany*,[35] for example, is an interview with Mendel Schainfeld, a German Jew who was sent to a death camp in Poland during the war, survived and emigrated to Norway. During a trip to Germany, he told his story to a filmmaker, who was content with presenting him, asking for further information, and adding a few images taken in contemporary Germany.

Historians will take into account the 'discourse' of the film and the bias that voices introduce in a movie. Diction is of paramount importance: an assertive speech sounds reliable, a faltering one seems less dependable. It must also be remembered that filmmakers make personal choices, even in the most 'neutral' documentaries. The director of *Mendel Schainfeld's Second Journey to Germany* was cautious not to get in the way of the interviewee, but he decided to film him in a train, to show that, now, the man was only a passenger in his former country. The effect would have been different if the meeting had taken place in a street or an office. Shooting is not a dispassionate activity. Director and operator never forget that audiences are sensitive to the quality of the pictures, so the second step in the analysis of a documentary will be a valuation of the cinematic work.

Is there an emphasis on the aesthetic? Cinema is not merely a recording of the world; it is able to express the pulse of life through formal and rhythmic combinations of pictures. Between the world wars, numerous filmmakers attempted to screen the 'feel' of big cities and represent their dynamic. The hectic pace of machines and rushing traffic, the combination of visually akin shapes, the tempo of montage, the patterns of images give these 'city poems' their indisputable charm. All pictures, shot in actual location, were genuine, but the play on time, the superimpositions, the parallels between mobiles built up a fascinating and totally abstract urban atmosphere. The most famous movie of this genre, *Berlin*, was not a report on the German capital but, as indicated by its subtitle, *The Symphony of a Great City*.[36]

The city films did not try to depict urban reality; they documented a perception, an idea of the flavour of towns. Other filmmakers, less interested in visual rhetoric, recorded the banality of everyday places and situations. *Neighbourhood 15*[37] explored, with apparent detachment, the borough of West Ham, presented the history and current condition of the area, exposed the problems people had to face, and took a look at the future. It offered an objective visit, but was carefully shot, with images taken right in the sun, well framed, astutely focused on smiling members of the community so that West Ham seemed a very pleasant place. What would it have looked like in heavy rain? Of course, style matters in written texts, but historians, who are used to taking care with literary devices, may feel disconcerted by cinematic tricks. For that reason, when a sequence appeals to them, they must carefully examine its structure, its framing (what was left outside?) and its lighting.

When everything has been scrutinized, what will be the next step? Films are neither texts nor photographs; they are moving images. Frozen or transcribed in words, they have lost their specificity; in fact they are reduced to nothingness. We can ponder an image or a phrase over the course of hours, but a movie exists only throughout time. The recourse to documentaries in historical research implies the introduction of moving pictures, of the flow of time in history works. This does not result in the ending of writing, for images need commentaries because they are unable to express theoretical notions such as causality, succession or anticipation, and cannot account for complex situations. Pictures introduce viewers directly to aspects of the past, while words help them stand back from immediate impressions. Combined films and texts are likely to modify our understanding of history. Documentaries, however, are human creations oriented by the personality of those who shot them. The choice and concatenation of images is subjective; we are not obliged to accept them as they are. Historians learned how to express themselves in phrases; today they must get used to selecting and editing pictures.

Notes

1 Notably the excellent Ian Aitken, ed., *Dictionary of the Documentary Film*, 3 vols (New York: Routledge, 2006) and Peter Zimmermann, ed., *Geschichte des Dokumentarischen Films in Deutschland, 1895–1945*, 3 vols (Stuttgart, Germany: Reclam, 2005).

2 On average an hour of fiction costs five times as much as an hour of documentary. Fictions can be broadcast twice, documentaries up to five times.

3 William Guynn, *A Cinema of Non-Fiction* (Cranbury, NJ: Associated University Press, 1990) assumes that documentaries can be characterized by contrast with fictions. Bill Nichols, *Representing Reality: Issues and Concepts in Documentary* (Bloomington, IN: Indiana University Press, 1991) and John Corner, *The Art of Record: A Critical Introduction to Documentary* (Manchester: Manchester University Press, 1996) are more cautious and find it difficult to give a definition.

4 *Der Weltkrieg*, 1927, 50-minute German compilation directed by Leo Lasko for UFA.

5 *Verdun: Visions d'histoire*, 1928, director Léon Poirier.

6 *The Great War*, 1964, BBC Television.

7 *World at War*, 1973, Thames Television.

8 *People's Century* was a 26-episode series co-produced by the BBC and the American Public Broadcasting Service put on air by the BBC in 1995.

9 *Douaumont*, 1931, director Heinz Paul.

10 *Nuit et brouillard*, 1955, director Alain Resnais.

11 *Gjutarna, the Iron Founders*, 1990, director Jean Hermanson.

12 *A Day in the Life of a Coal Miner*, 1910, unknown director. Produced by Kineto, a subsidiary of the Charles Urban Trading Company, founded as a documentary maker in 1909.

13 *Coal Face*, 1935, director Alberto Cavalcanti. Produced by the GPO Film Unit.

14 *Living on the Edge*, broadcast on Channel Four, November 1987, director Michael Grigsby.

15 *A Complaint of Rape*, broadcast by the BBC in 1982, director Roger Graef.

16 *Health of the Nation*, 1937, director Charles Barnett.

17 *Drifters*, 1929, director John Grierson.

18 *Nanook of the North*, 1922, director Robert Flaherty.

19 *Der VW Komplex*, 1989, director Hartmut Bitomsky.

20 *The Thin Blue Line*, 1988, director Errol Morris.

21 *Las Hurdes: Tierra sin pan*, 1936, director Luis Buñuel.

22 *Fishing Party*, broadcast on BBC Two, 27 February 1986, director Paul Watson.

23 *Death on the Rock*, broadcast on ITV, 28 April 1988, director Roger Bolton.

24 *Song of Ceylon*, 1934, director Basil Wright.

25 *El enemigo principal*, 1974, produced by the revolutionary Ukamau Film Collective, director Jorge Sanjinés.

26 *Wake Island*, 1942, where faked Japanese massacre defenceless civilians.

27 *Triumph des Willens*, 1934, and *Olympia*, 1936, both directed by Leni Riefenstahl.

28 *Der ewige Jude*, 1940, director Fritz Hippler.

29 *General'naya liniya*, 1929, director Sergei Eisenstein.

30 *La hora de los hornos*, 1968, directors Fernando Solanas and Octavio Getiño.

31 *Blood of the Condor*, 1969, produced by the Ukamau Film Collective, director Jorge Sanjinés.

32   *Victory at Sea* was a 26-episode programme broadcast by NBC in 1952.
33   *Housing Problems*, 1935, directors Edgar Antsey and Arthur Elton. Produced by the British Gas and Coke Company.
34   *Primary*, 1959, director Richard Leacock.
35   *Mendel Schainfelds zweite Reise nach Deutschland*, television documentary directed by Hans-Dieter Grabe, broadcast on the German ZDF channel on 11 March 1972.
36   *Berlin: die Sinfonie der Grosstadt*, 1927, director Walter Ruttmann.
37   *Neighbourhood 15*, 1948, produced by the Land and Learn Film Unit.

8 More than just entertainment: the feature film and the historian

by Michael Paris

The birth of cinema

By the late nineteenth century a number of inventors were attempting to develop a means of recording and projecting the moving image, but in the early 1890s it was the American inventor Thomas Edison who created a viable commercial system. With the 'Kinetograph' camera that could record moving pictures and the 'Kinetoscope', an individual peepshow through which that 'motion' picture could be viewed, Edison took the steps that resulted in his receiving credit as the creator of the movies. Well aware of the commercial potential of this development, in April 1894 he rented a shop and opened a Kinetoscope Parlour on New York's Broadway; the age of cinema was at hand. But Edison was not alone in his desire to record and project moving pictures. Throughout Europe many individuals were infused with the same ambition, but it was in France that the next step was made. The Lumière Brothers, Auguste and Louis, developed their own camera, the *cinematographe*, a more portable version of Edison's invention, which doubled as a projector. In 1895, the Lumierès demonstrated their invention at a number of public film shows throughout France, and even in London. Film shows were an immediate sensation for the public and such entertainments became a novel attraction at fairs and as a feature in vaudeville shows.

The first films were simply short actuality sequences – trains leaving or arriving at a station, busy city streets, workers leaving a factory – but they were enough to enthral audiences and convince them they were witnessing reality. Filmmakers soon became more ambitious and began to record the great events of the age: the funeral of Queen Victoria, British troops in action during the Boer War or episodes from the Spanish American conflict. When the camera was unavailable to record these dramatic happenings, the filmmakers simply recreated them in the studio using models and a variety of photographic trickery. Biograph's film of the 1906 San Francisco earthquake convinced many that the camera had really been witness to the tragedy. Here, then, the filmmaker was already acting as historian, recording or recreating and manipulating great events for their audience. But the public were fickle, and despite the great novelty value of the moving image, audiences soon grew bored with the same shots, as Edison had in fact predicted. Filmmakers, aware of the transient nature of their audiences, looked for different forms in which the moving picture could be employed. Documentary films and newsreels, a weekly or bi-weekly review of the great news stories of the day, were soon an established part of the great picture show. But the really significant development was the conception of the narrative film – the fictional story, a melodramatic or comedic sequence, a chase, a narrative of heroism, and so on; and it was the introduction of narrative that, as Arthur Knight has argued, saved the movies.[1] As the novelty value of the moving picture faded, audiences would soon have lost interest in actuality sequences however sensational, but the emergence of filmic narratives brought a whole new dimension to cinema and transformed a mechanical novelty into a new art form: a form that was easily accessible and opened almost endless possibilities for storytelling. Georges Melies' exciting fantasy *A Trip to the Moon* and Edwin S. Porter's *The Life of an American Fireman* (both 1902) entranced audiences and paved the way for the new art form.[2] This chapter, then, is concerned with the narrative, the fiction film, and what it can offer to historians as a dynamic and vital source for the social and political history of the twentieth century (Fig. 8.1).

In the early twentieth century, films became enormously popular, especially with the working classes, and film shows moved out from the fairgrounds and variety theatres, into purpose-built cinemas whose business was screening movies. In an effort to attract a better class of clientele, filmmakers turned to the legitimate theatre for inspiration and film versions of classic and modern plays found their way onto film. This transition was not always successful, but it did result in longer, and generally better, movies. The historical film, for example, thrived in Italy, where filmmakers who were anxious to help legitimize the claims of their nation to world power status looked back to the splendours of the Roman past in epic productions like the 1912 version of *Quo Vadis*, which ran for over two hours. Such efforts inspired other filmmakers with the exciting possibilities of using film to recreate dramatic episodes from the past. By the outbreak of the First World War, film had become established as potentially one of the most significant and popular art forms, but the experience of 1914–18 was to further transform cinema.

The First World War had considerable impact on the movies, creating an enormous market for films, which were one dimension of the demand for popular entertainment which would, if only for an hour or two, provide audiences with a distraction from the dangers and hardships of life in wartime. The cinema had never been so popular, and the involvement of government agencies in film production in an attempt to use the medium as an agency for mass persuasion provided much needed respectability for film. But if the war acted as a forcing house for the development of national cinemas, the closed frontiers of Europe enabled the US film industry to establish a dominance in the production and distribution of films that has never been seriously challenged.

The social habit of the age

From the 1920s to the late 1950s cinema was the predominant leisure activity for millions – a worldwide phenomenon that, as historian A.J.P. Taylor famously noted, became 'the essential social habit of the age'.[3] The United States led the way: by 1939, 80 million tickets were sold every week, and throughout Europe films were equally popular. But the feature film was more than just a diverting entertainment: it was also an important source of information, a vital element in helping to shape the worldview of the audience. Film interpreted great events, made sense of both the modern world and the past, and taught all manner of useful lessons. Lifelong film enthusiast and television executive

Figure 8.1 'Cinématographie Perfectionné'

Leslie Halliwell recalled in his autobiography some of the many things he had learned from 'the pictures' as an adolescent:

> Real life was fascinating, but untidy and sometimes sad; the kind of life shown on the silver screen had dramatic progression, and its loose ends of plot were always tied up. ... It was highly moral, and taught me such things as how to behave at table, how to speak to a lady, and what was involved in various kinds of adult activity. ... It gave me an idea of what happened in history. ... It gave me a taste for the rhythms of popular music.[4]

The feature film is essentially an audiovisual narrative, but the real source of its popularity is its capacity to draw viewers into the unfolding story on the screen and enable them to identify with the characters and, in a sense, to share their experiences. Clearly it is this aspect of film, what we might call the 'magic of cinema' that accounts for its enormous impact and influence upon the viewer. Here, for a short time, we share the dangers of war, the hardships of trekking through the desert, or the heartache of a domestic tragedy. We can share intimate moments with the glamorous or the comedic aspects of life. We know what we see on the screen is not real, but for an hour or two we suspend disbelief and enter wholeheartedly into that world of make-believe, and when we leave the auditorium some of that experience remains to shape how we react and think in similar circumstances. The magic of cinema is indeed a powerful force – and it is that factor which makes film such an effective method of persuasion and social control. The transmission of the dominant ideas, attitudes and values of the culture in which the film is made underlie its entertainment function and make it a powerful means of mass persuasion, as it both reflects and reinforces popular ideas and preoccupations and inculcates views and attitudes deemed desirable by the filmmaker and wider society.

Politics and film

The potential of film as a political weapon was first demonstrated during the First World War. That conflict was the first total war, one in which every national resource was eventually called into play. By 1915 the governments of the combatant nations were beginning to utilize film as part of their historical record. But film soon developed a propaganda function that justified the nation's war effort, ridiculed and demonized the enemy, heroicized the armed forces, and encouraged the home front to ever greater effort. Propaganda may be defined as the attempt to influence public opinion through the transmission of particular ideas and values. The term often evokes the lies and distortions employed by the despotic regimes to strengthen their hold over a populace, but in its softer version it is also employed in democratic societies by both government and businesses. Short information films were soon supplemented by animated cartoons, feature-length documentaries, such as the British War Office's epic *The Battle of the Somme* (1916) and lavish feature narratives such as D.W. Griffith's ill-fated epic *Hearts of the World* (1918), which was intended to arouse the Allies to a new level of hatred against the Hun. By the end of the war British propaganda (including film) was judged to have been so effective that in the United States it was widely believed that British propaganda had been primarily responsible for American entry into the war in 1917.[5]

The interaction of politics and film was to be most clearly seen in Russia after the Bolshevik Revolution of 1917. 'Of all the arts, for us cinema is the most important,' proclaimed Lenin, and indeed film quickly became one of the principal agencies for the dissemination of Soviet propaganda. Film is a visual medium and in the years before the introduction of sound, it was purely visual. Thus it seemed an ideal medium through which to influence the predominantly illiterate, unsophisticated peasantry of Russia. Moreover, films had universal appeal, for what happens on the screen is easily under-stood by all, whether educated or illiterate, and operates on both the emotions of the individual and the group mind. By 1919, the Bolsheviks had taken control of the film industry and centralized control under the People's Commissariat for Enlightenment. New, hastily created cinemas and the mobile 'agitprop' trains, equipped for showing films and with their own team of propagandists, took the Soviet message to even the most remote areas of that vast country, while theorists like Lev Kuhlesov investigated the ways in the moving visual images communicate.

But Lenin was shrewd enough to realize that a filmic diet of unrelenting propaganda would be counterproductive and he urged the production of straightforward entertain-ment films as well. These would attract audiences into the cinemas and produce the revenue necessary to create a thriving film industry. Once the habit of filmgoing was established, audiences would be available for exposure to political messages as well. Unfortunately, from this point of view, intellectual Soviet filmmakers of the 1920s, like Sergei Eisenstein and Vsevolod I. Pudovkin, failed to take sufficient heed of Lenin's advice and, although their great propaganda films such as *Battleship Potemkin*, *October* and *Mother* were much admired by party intellectuals and art-house audiences in the west, the Russian peasantry found them obscure and dull, much preferring a Chaplin comedy or a Mary Pickford melodrama. The problem was addressed in 1928 when the decision was made to make film more intelligible to the masses. The development of a political cinema in Russia was to have considerable influence on other totalitarian regimes (Fig.8.2).[6]

In Fascist Italy and especially in Nazi Germany, film propaganda was seen to be a key element in the regime's efforts of internal social control and propaganda to the outside world. Both took almost immediate control of the mass media for these purposes. In Germany, Joseph Goebbels, himself a film enthusiast, exercised rigid control over the film industry, arguably the most powerful in Europe. Through his Ministry for Popular Enlightenment and Propaganda, German film was made into an agent for drumming up active support for the New Order. Non-Aryans were excluded from the production pro-cess and direct film propaganda was confined to the newsreel or to elaborate docu-mentaries such as *Triumph of the Will* (1935) and *Victory in the West* (1941). Feature films were also employed to reinforce the political views of the regime: *Jew Suss* and *The Rothschilds* (both 1940), for instance, supported anti-Semitic policies, while *Uncle Kruger* or *My Life for Ireland* (both 1941) stoked anti-British sentiments.[7]

But it was not only in totalitarian states where feature films were employed by the propagandist. In Britain, where the government claimed publicly to eschew propaganda, film was used by the political parties and in a broader sense to fly the flag for the Empire, for British manufacturing and for the British way. But British studios, in a state of almost continual financial crisis, simply could not compete with the high production values of

Figure 8.2 Alexander Rodchenko's striking poster

American studios and the glamour of American 'stars'. In 1937, for example, John Grierson, leader of the fledgling British documentary movement, despaired that 'so far as films go, we are now a colonial people'.[8] The domination of British screens by Hollywood products and the continued failure of British films to find an audience, even at home, frequently provoked critical debate about how the industry could be sustained; and not simply because of the need to preserve a significant element of the economy, but because film was also a vitally important channel for the dissemination of national propaganda. Stephen Tallents, the civil servant most responsible for the organization of government film propaganda in the late 1920s, shrewdly observed that, because of the economic disruption of the First World War and the economic collapse of the early 1930s, a positive representation of the nation and empire in the media, and especially in film, was essential for national recovery:

> If we are to play our part in the New World order, we need to muster every means and every art by which we can communicate with other Peoples. ... We need continuous and sustained presentation of our industrial ability and industrial ambitions.[9]

Cinema, then, had a significant role to play in national life and in marketing the nation and its products overseas. Attracting larger audiences than any other form of entertainment, it urgently required British films that promoted British interests. In short, what was needed was a distinctively 'national' cinema. The Moyne Committee on the future of British film recognized this point in 1936, when it reported:

The cinematograph today is one of the most widely used means for the amusement of the public at large. It is also undoubtedly a most important factor in the education of all classes of the community, in the spread of national culture and in presenting national ideas and customs to the world. Its potentialities moreover in shaping the ideas of the very large numbers to whom it appeals are almost unlimited. The propaganda value of film cannot be overemphasized.[10]

In the United States, feature film had always reflected the 'American way' and, under the control of corporations sometimes motivated by fear of government intervention or social control and sometimes mainly by market factors, had broadly supported the dominant attitudes, values and political consensus. In the years after the Bolshevik revolution in Russia, filmmakers had reflected and reinforced official policy towards Russia and had made much of the danger posed by communism and the superiority of democracy and laissez-faire capitalism. Then, during the Great Depression, cinema played its part in providing escapist entertainment, calming fears and demonstrating that American 'get up and go' would eventually win through. At the same time, there was steady technical improvement, with the arrival of sound in the late 1920s, colour in the 1930s and generally better quality and production values. Thus when the Second World War came, the movie industries of the combatant nations were willing conscripts for the duration, producing a vast array of popular feature films which supported the war effort, revealed the evil nature of the enemy and translated war aims into easily understood stories with which audiences could identify. Not that all films were heavy with political content of course, but even the musicals, comedies and escapist fantasies performed the political function of maintaining public morale in a time of crisis. As the post-war world divided into super-power rivalry, national film industries continued to offer covert, and sometimes more obvious, political support for dominant political ideologies by reflecting the social mores and attitudes in a rapidly changing world.

Film and the historian

Since the birth of cinema at the end of the nineteenth century, filmmakers have been recording and interpreting the world around them, reflecting the social and political realities of the society in which they worked. Most of the great events of the twentieth century have been captured on film: wars, revolutions, disasters both human-made and natural, important sporting events, political leaders and the great and noted personalities from every field of endeavour. And those significant moments when the camera was not present have been recreated later in the studio. Since the First World War, film has also been employed as an agent of mass persuasion by governments anxious to shape popular attitudes and to gain support for their policies. Films, therefore, are a powerful window on the past – a source that often allows us not only to see the actual event, or its restaging, but also to understand how the society that produced the film wanted its audience to respond. The technical improvements introduced during and after the 1930s – sound, colour and so on – increased audience understanding and made more complex narratives possible, but they also added further layers of meaning to the film. During the first half of the twentieth century, however, historians had little time for this vast resource, and it was not until the later 1960s that a handful of scholars tentatively began to investigate the possibilities offered by the moving image, and thus began the 'film and history' movement.

Some historians had already begun to focus upon the cinema as a social phenomenon, in much the same way they had investigated other key institutions of modern society, but histories of various national film industries were notoriously unreliable – often little more than a record of technological developments, box-office successes and gossip about the private lives of the movie moguls and popular stars. Not until 1949 did Rachael Low embark upon her monumental academic institutional and cultural history of the British film industry. This marked the birth of scholarly interest in the cinema, but there was still considerable reluctance to mine films for evidence of the social and political history of the twentieth century in the same way that academics were prepared to use popular fiction, drama and painting. In part, this long neglect of film can be explained by the inherent conservatism of the historical profession and its reliance on the written word as the most valuable form of evidence, but it was also partly due to the nature of film itself. As John O'Connor, an early advocate of the documentary importance of visual images pointed out, as late as the 1970s, it was difficult to make use of films, because

> There is no other area of research where the historian may be forced to rent his text from a commercial distributor and pay for each screening, and it is difficult (without special equipment) to slow down or stop the film or to go over a section again as one can with a printed document.[11]

The development of the video recorder gradually provided for easier access to films but considerable problems remained in locating the desired films and the necessary back-up documents on production, distribution and reception. Above all, there was no methodology, and the widely held conviction that film was somehow not respectable, or constituted at best a dubious historical source, proved to be durable. Nevertheless, by the 1960s film was being used in classroom teaching, albeit as incidental illustration, and historians such as Rolf Shuursma in the Netherlands and John Grenville and Nicholas Pronay in Britain were beginning to make films for classroom use. The Universities Film Committee was set up in Britain in 1968, and in 1970 the American Historical Association established the Historians Film Committee and its journal *Film and History*, both of which encouraged scholars to include film material in the accepted body of historical source material. The movies were 'suspended somewhere between the hell of mass culture and the heaven of high art', W.R. Robinson observed around that time. 'They are undergoing aesthetic purification, with the favourably disposed intellectuals as their advocates and the university as their purgatory.'[12]

It was not surprising that the first attempts to use filmic evidence were drawn from the 'actuality footage' of newsreels and documentaries. As Anthony Aldgate and Jeffrey Richards have noted, historians confronting these products were 'reassured by the presence of real people and real locations that they were somehow viewing reality'.[13] These forms of films, perhaps, seemed closer to the traditional source materials and were as well films of record, offering visual evidence of social and political events but shaped or interpreted in line with the official view. Nicholas Pronay's work on British newsreels of the 1930s and Aldgate's study of how the Spanish Civil War was reported by newsreel were important landmarks in opening up a rich vein of filmic sources.[14]

These careful works made clear that, far from being simple entertainment, the feature film also carries a great deal of social and political information that can assist the historian in unravelling the past. In the late 1960s the Swedish academics Leif Furhammar and Folke Isaksson produced their ground-breaking study, now sadly neglected, *Politics and Film*. In a broad survey of popular films from several countries, the authors presented a 'collection of essays on movies which have a clear political purpose, on cinema as a weapon of propaganda'.[15] The authors did not simply focus on the more obvious examples of film used for political purposes, during wars or by totalitarian regimes, but included a section on the ways popular Hollywood films have supported dominant values and attitudes and interpreted the wider world for the American audience. Their ideas were taken forward in a number of studies that appeared over the next few years, in particular Jeffrey Richards' masterful *Visions of Yesterday*, published in 1973. Richards offered three case studies – Nazi Cinema, American populism and the British cinema of Empire – each study carefully constructed to demonstrate how filmmakers had absorbed the dominant ideologies and translated those ideas into popular film which both reflects and reinforces the values society deems worthwhile preserving.[16]

The practicalities of using film as an historical source were examined in a variety of ways through the essays in *The Historian and Film*, edited by Paul Smith in 1976. The essays ranged widely from the preservation of film, locating archives, using film in the classroom, and the historian as filmmaker to several studies on the use of film archives and on the evaluation of film as historical evidence.[17] Unfortunately, as Pierre Sorlin has noted, these essays offer 'a great deal of information on collecting, restoring, criticizing and editing film material', but pay 'scant attention' to fiction films.[18] This omission was rectified in the early 1980s by a number of works that advanced the study of the feature film, particularly K.R.M. Short's *The Feature Film and History*, a stimulating collection of essays by some of the leading practitioners in the field.[19] Here the contributors examined a number of important American and European mid-twentieth-century themes using feature films as their primary source. Their subjects included the developing national consciousness in 1920s France and Germany, American liberalism, the search for a national consensus in 1930s Britain, and fighting anti-Semitism in 1940s America. These seminal works all developed and refined a practical methodology for using filmic evidence making use of the authors' own research and drawing upon other recent work in the field.

These studies helped to develop a methodology for using film based upon contextual analysis: a close reading of the film itself, the context of production and its reception. The feature film contains a great deal of information, not just its story (what we might call the 'witting testimony') but also background information (the 'unwitting testimony'). Films made during the 1930s, for example, illustrate the background of life during that decade, the style of the times, housing, streets, fashions, transportation, social conventions and so on – all useful visual evidence for getting a sense of the times and illustrative of the social background. The narrative, whether a drama, love story or whatever, will be acted out by characters who voice the prevailing attitudes, values and social mores of the time, its hopes, fears and concerns. All films, then, can tell us something about the society in which they are produced but this becomes more valuable to the historian if the film deals with a social problem or political issue. The popular American gangster cycle of the 1930s, for example, reflected the concern of many at the growing levels of violence and the corruption within American institutions due to the rise of organized crime. Some films documented this phenomenon and attempted to explain it through the psycho-pathology of the gangster – *Little Caesar* (Mervyn Le Roy, 1930), *Public Enemy* (William Wellman, 1931) and *Scarface* (Howard Hawks, 1931). Others attempted to find a more satisfying and more logical explanation in the poverty and sheer hopelessness caused by the Depression, for example *Dead End* (William Wyler, 1938) and *The Roaring Twenties* (Raoul Walsh, 1939).[20]

Films made during the Second World War, when all the combatant nations conscripted film as a significant element in their war effort, tell us a great deal about how governments wanted their citizens to think about what they were fighting for and how to regard the enemy and their Allies. An interesting example is the pro-Soviet films made after 1942 when the United States and Soviet Russia were shoulder to shoulder in the struggle against Nazism. Hollywood, traditionally uncompromisingly anti-communist, was suddenly faced with selling Russia as a loyal and noble ally to Middle America. The result was a spate of films that portrayed the Russians as decent, ordinary people, driven by the same concerns and anxieties as Americans. They cared about their towns, their parents, their children and they wanted to be free to decide their own future. Essentially the inhabitants of *North Star* (Lewis Milestone, 1943), a small rural village invaded by the Nazis, react in exactly the same way any small town Americans would to a similar danger. These people are not Bolsheviks bent upon world revolution, but decent folk doing the best they can and just like us. Films like *North Star* did their work well persuading many Americans that they had little to fear from the Russians, but the irony was that after 1945 and the collapse of the wartime grand alliance, Russia was again demonized and Hollywood, responding to political pressure, became the mouthpiece for extreme political views in a stream of Cold War movies that preached the dangers of the Soviet system.[21]

But a close reading of the film itself needs to be extended beyond a mere surface analysis. As John O'Connor argues: 'the cultural analysis of film demands that attention must be given to the ways in which the artefact was understood by historical spectators at the time of its production and release'.[22] Because meanings, attitudes and values change over time, there is clearly a 'culture gap' between those who made and viewed the film at the time of its original release and those who view it later as a historical artefact. Thus, we need to think ourselves back into the mindset of an audience in the 1930s, or to whatever decade our film was made, and the closer we get to 'getting inside the skin' of that viewer, the more we will learn from that film. An understanding of the historical context is therefore essential to a proper understanding of the film being studied, and we can go still deeper by exploring how and why the film was made, by pursuing what Thomas Cripps has called 'the paper trail' of production.[23]

Whatever the source, there are basic questions that the historian must ask: why was this source produced, what was its purpose, who created it, and so on; and a filmic source is no exception. Before we can effectively use a film we need to know who made it and why. Yet with collaborative projects like film, these questions are not easy. We must consider the writers, director, producer, perhaps even the actors who may have had some creative input to the finished product, as well as the influence of those who financed the picture. It is clearly this collaborative aspect that gives film its historical value as a representation of popular attitudes, but which, at the same time, makes its production history complex and difficult to research. Furthermore, until recently, most film scholars have not been historians. An emphasis on the textual analysis of film by those trained in literature, art or philosophy is hardly surprising and, in any case, historians have reaped benefits from the prevailing emphasis on film-as-text. Still, historians entering the realm of film studies have often been uncomfortable with the ahistorical nature of much film analysis. But since the early 1990s, an increasingly recognizable group of scholars has worked energetically to develop a more soundly historical approach. Film history, they argue, must be more than only the analysis of *films*; it must examine the whole context of cinema, including both the process of film*making* and the experience of film*going*. Seeking to correct what one of them has described as 'our appalling ignorance' of some of the most basic aspects of the environments in which films are made, many of these researchers have focused on subjects such as the film studios as business organizations; the distribution and exhibition of films; and the legal, political and economic considerations that have played their parts in determining what audiences would find in their local cinemas.[24] Films are commercial properties: to succeed they must attract an audience, and filmmakers are constantly aware of what might be termed their 'target audience', the groups to which the film will appeal. It can be assumed, then, that the most financially successful films are those which appeal to the broadest audience, and that filmmakers will inevitably go out of their way to appeal to all possible audiences. The old maxim that controversy meant death at the box-office held sway in Hollywood for decades, and it is precisely because of it that film tended to reflect and reinforce genuine mass opinion and, consequently, to have value as an indicator of social and political attitudes.[25]

Closely allied to production history is the question of reception: who saw the film, how was it received at the time of its release and what effect might it have had upon audiences? 'Historical reception studies', which seek answers to questions of this kind, confront some daunting theoretical and methodological challenges, for it is difficult to discern the opinions and reactions of contemporary audiences, let alone those of earlier times. Not surprisingly, the Hollywood studios have long encouraged belief in the near impossibility of this undertaking, for it supports their preference to be thought of as 'dream factories', engaged in the commercially risky creation of artistic products whose success or failure can hardly be foreseen. Traditional film studies offered little challenge to Hollywood's self-portrait, but the recent scholarship has asserted that the studios are in fact effectively managed business corporations that are far from ignorant of the preferences of their potential customers.[26] Therefore, the studio archives, to the extent they are open, are useful for historical reception studies, but they must be examined in combination with local resources, such as old newspapers, city directories and museum collections, because film audiences must be studied, as it were, *in situ*. The size of the task and the scarcity and scattering of resources might suggest that little could be learned from this research. But the proponents of historical reception studies have proven themselves to be extremely imaginative. In addition to the traditional research methods mentioned above, they have employed tools such as oral history, historiographical adaptations of the techniques of audience surveying, such as focus groups, ethnographic methods (sometimes called 'memory work'), and statistical analysis of business records and other data. Given the long domination of film studies by 'high theory', it is interesting to observe that the new film history has paralleled the rise of post-structuralist literary theory, which removed text from its position of centrality, particularly the work of critics such as Hans Robert Jauss, Stanley Fish and Wolfgang Iser.

> Central to the reading of every literary work [Iser argues] is the interaction between its structure and its recipient. This is why the phenomenological study of art has emphatically drawn attention to the fact that the study of a literary work should concern not only the actual text but also, and in equal measure, the actions involved in responding to the text.[27]

The substitution of 'film' for 'literary work' would make Iser's statement equally applicable to historical reception studies.

The historical film

Films, then, can offer the historian a mirror for the concerns, anxieties and values of the society in which they are made, but what of historical films – the filmmakers' attempt to recreate the past? While any definition of the 'historical film' will be contentious, for the purposes of this chapter let us say that a historical film is one in which the narrative is set wholly or partly in the past and is based upon real events and actual people. A second category of historical films consists of wholly fictional stories set in the past – westerns, medieval romps or melodramas set in some vague, sometimes ill-defined historical period – such as *Wagonmaster* (1950), *The Black Shield of Falworth* (1954) or *The Wicked Lady* (1945). Although these films make use of historical settings, their essential effect is simply to exploit the romance, adventure and pageantry of the past. They are perhaps best described as 'costume dramas', for they often have little to do with history as we know it. In both cases, however, filmmakers will have few qualms about sacrificing historical accuracy for good entertainment because a film is a commercial enterprise seeking maximum profitability. Thus, invented characters or episodes, or distortions of truth can be used to enhance the drama of the narrative; to a more satisfying and more commercially successful, drama. Oliver Stone's much discussed film about the Kennedy assassination, *JFK* (1991), is a useful case in point. In order to heighten the drama, to provoke new discussion, and to promote his own view of America as military-industrial complex, Stone invented characters, episodes and even manufactured seemingly genuine newsreel footage to advance his thesis. Clearly not all historical films are subject to such extreme interpretations, but they always reflect the filmmakers' own views of what happened and why it happened, and must be tailored to the dictates of the box-office.

While good, accurate historical reconstructions on film can remind a mass audience of episodes and events that have been largely forgotten (*Glory* (1987) and *The Tuskegee Airmen* (1992) for example, are, arguably, timely reminders of the contribution of African Americans to the United States' wars), or provide a useful introduction to an event or the contribution of a significant historical figure, we cannot expect the same degree of historical accuracy from the cinematic text as from a carefully researched monograph. It would be foolish to assume that a film like Errol Flynn's swashbuckling masterpiece

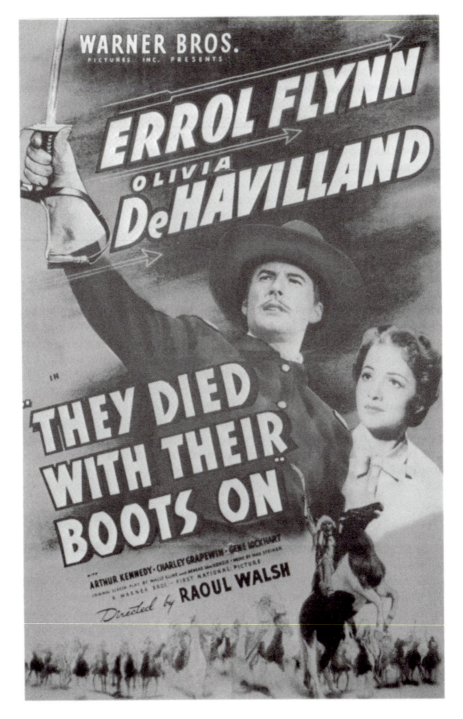

Figure 8.3 Warner Bros take on the Custer legend

The Adventures of Robin Hood (1938) will, however enjoyable, tell us anything useful about the state of England in the twelfth century. So are historical films of any value at all to the serious historian? Some scholars advocate ignoring films that are essentially attempts to reconstruct historical events, while others, most notably Robert Rosenstone, argue that film is a unique medium that teaches about the past in its own way and must be understood on its own terms.

In either case, it is important to recognize that historical films usually have more to say about the present in which they were made than about the past in which they are set. Using an historical parallel is a relatively 'safe' way for a filmmaker to comment on his or her own period. During the Vietnam War, for example, even though it might have been considered unAmerican to criticize the war, some filmmakers were able to offer a critique obliquely through the Westerns or other genres. *Chato's Land* (1971), a brutal story in which racist white Americans pursue a half-breed Apache into 'Indian territory' and to their deaths is an obvious example. Similarly, Eisenstein's epic *Alexander Nevsky* (1938), which deals with an invasion of Russia by the German Knights of the Teutonic Order, is clearly indicative of the deep distrust which many Russians felt for Hitler's Germany at that time; while Korda's *Lady Hamilton* (1941), set against the background of Britain's successful struggle with Napoleon's France, was clearly an allegory whose meaning would have been quite obvious to Anglo-American audiences of the time.[28] Many filmmakers have plundered the past in order to find subject matter appropriate for lessons for their own society. Historical films, then, can offer insights into the problems of our own societies by exploring the past, and, as noted above, a well-made historical film not only can stimulate new public interest and debate about a period, event or historical character, but also can sometimes offer a sense of the past that is impossible to gain from the printed word (Fig. 8.3).

Since the beginning of the twentieth century, the cinema has recorded and interpreted our social and political life and has made those interpretations available to a mass audience. Historians simply cannot afford to ignore what film tells us about our society and our past.

Notes

1 Arthur Knight, *The Liveliest Art: A Panoramic History of the Movies* (New York: Mentor, 1979), 14.

2 On the history of cinema, see Rachael Low, *The History of the British Film, 1896–1939*, 7 vols (London: Allen & Unwin, 1949–85); Robert Sklar, *Movie-Made America: A Cultural History of American Movies* (New York: Vintage, 1976). For a broad international history, see James Chapman, *Cinemas of the World: Film and Society from 1895 to the Present* (London: Reaktion, 2003).

3 A.J.P. Taylor, *English History, 1914–1945* (Oxford: Oxford University Press, 1965), 313.

4 Leslie Halliwell, *Seats in All Parts: Half a Lifetime at the Movies* (London: Grafton, 1986), 25.

5 On cinema and the First World War, see Michael T. Isenberg, *War on Film: American Cinema and World War 1* (London: Associated University Presses, 1981); Leslie Midkiff Debauche, *Reel Patriotism: The Movies and World War One* (Madison, WI: University of Wisconsin Press, 1997); Andrew Kelly, *Cinema and the Great War* (London: Routledge,

1997); Michael Paris, ed., *The First World War and Popular Cinema* (Edinburgh: Edinburgh University Press, 1999); Bernadette Kester, *Film Front Weimar: Representations of the First World War in German Films of the Weimar Period, 1919–1933* (Amsterdam: Amsterdam University Press, 2003).

6   For an excellent introduction to the Soviet use of feature film, see Richard Taylor, *Film Propaganda: Soviet Russia and Nazi Germany* (London: Tauris, 1988).

7   Nazi cinema is dealt with in David Welch, *Propaganda and the German Cinema, 1933–1945* (Oxford: Oxford University Press, 1990) and *The Third Reich: Politics and Propaganda* (London: Routledge, 1993).

8   John Grierson, *World Film News* 2, no. 8 (1937): 5.

9   Stephen Tallents, *The Projection of England* (London: Faber & Faber, 1932), 17.

10   Quoted in Anthony Aldgate and Jeffrey Richards, *Best of British: Cinema and Society from 1930 to the Present* (London: Tauris, 1999), 1–2.

11   John E. O'Connor and Martin A. Jackson, eds., *American History/American Film* (New York: Frederick Ungar, 1979), xv–xvi.

12   W.R. Robinson, ed., *Man and the Movies* (Baltimore, MD: Penguin, 1867), 6.

13   Aldgate and Richards, *Best of British*, 4.

14   Nicholas Pronay, 'British Newsreels in the 1930's 1: Audience and Producers', *History* 56, no. 188 (1971): 411–18; 'British Newsreels in the 1930's 2: Their Policies and Impact', *History* 57, no. 189 (1972: 63–72; Anthony Aldgate, *Cinema and History: British Newsreels and the Spanish Civil War* (London: Scolar Press, 1979).

15   Leif Furhammar and Folke Isaksson, *Politics and Film* (London: Studio Vista, 1971), 6.

16   Jeffrey Richards, *Visions of Yesterday* (London: Routledge & Kegan Paul, 1973).

17   Paul Smith, ed., *The Historian and Film* (Cambridge: Cambridge University Press, 1976).

18   Pierre Sorlin, *The Film in History: Restaging the Past* (Oxford: Blackwell, 1980), vii.

19   K.R.M. Short, ed., *The Feature Film and History* (Knoxville, TN: University of Tennessee Press, 1981).

20   On gangster films, see Colin McArthur, *Underworld USA* (London: Secker & Warburg, 1972).

21   On pro-Soviet and Cold War movies see, for example, Daniel J. Leab, 'How Red Was My Valley: Hollywood, the Cold War and *I Married a Communist*', *Journal of Contemporary History* 19 (1984): 59–88.

22   John E. O'Connor, ed., *Image as Artifact: The Historical Analysis of Film and Television* (Malabar, FL: Krieger, 1990), 110.

23   Thomas Cripps, 'The Moving Image as Social History: Stalking the Paper Trail', in *Image as Artifact*, ed. O'Connor, 136–55.

24   Robert C. Allen, 'From Exhibition to Reception: Reflections on the Audience in Film History', in *Screen Histories: A Screen Reader*, ed. Annette Kuhn and Jackie Stacey (Oxford: Oxford University Press, 1998), 14.

25   See William Hughes, 'The Evaluation of Film as Evidence', in *The Historian and Film*, ed. Smith, 49–79.

26   Richard Maltby, 'Sticks, Hicks and Flaps: Classical Hollywood's Generic Conception of the Audience', in *Identifying Hollywood's Audiences: Cultural Identity and the Movies*, ed. Richard Maltby and Melvyn Stokes (London: British Film Institute, 1999), 23–4.

27   Wolfgang Iser, 'Interaction between Text and Reader', *The Norton Anthology of Theory and Criticism* (New York: W.W. Norton, 2001), 1671.

28 On American historical films, see O'Connor and Jackson, *American History/American Film*. For British examples, see James Chapman, *Past and Present: National Identity and the British Historical Film* (London: Tauris, 2005).

9   The visual culture of television news

by Cynthia Carter and Stuart Allan

Introduction

In the event that you have been fortunate enough to travel to various countries around the globe, it is possible that you will have been struck by the remarkable similarities in television news from one country to the next. More often than not, at least in our experience, the national evening newscast commences with an image of a revolving globe (the country in question is usually highlighted) to the sound of a sharply ascending piece of theme music. The opening sequence, having established an urgent sense of immediacy, promptly gives way to a shot of the studio, a pristine place of hard, polished surfaces (connotations of efficiency and objectivity) devoid of everyday, human (subjective) features. The camera glides smoothly across the studio floor, coming to a halt in front of the newsreader, who is seated squarely behind a desk, his or her gaze ostensibly affixed on us. He or she is formally attired, their body movements calmly measured, with a speaking voice (inevitably a 'proper' accent) solemn and resolute in tone. The ordering of the news stories to follow will typically respect a temporal rhythm, each self-contained item positioned in descending order of ascribed newsworthiness, until closure is invoked about 30 minutes later – but only after a final, 'human interest' story has provided a light-hearted observation or two about the foibles of daily life. In upholding these types of rules, the newscast offers the viewer a sense of continuity, not least by the sheer repetition of its narrative conventions on an everyday basis.

Variations on this format exist, of course, but it is likely to sound familiar. So familiar, perhaps, that it may be difficult to envisage alternative ways of constructing a newscast, so ostensibly natural is the alignment of these narrative conventions with connotations of credibility, authority and truthfulness – and thereby journalistic professionalism. Yet, there is nothing intrinsically *natural* about these conventions at all. That is to say, while newscasts typically seek to present a 'mirror' of society, with each news account an 'objective' or 'impartial' *translation* of reality, it is nonetheless possible to argue that these accounts are actually providing ideological *constructions* of contending truth-claims about reality. This is to suggest that the newscast, far from simply reflecting 'the world out there', is effectively setting in motion codified definitions – or frames – of what should count as the reality of the events being presented.[1]

Accordingly, an analysis of the visual culture of television news will need to discern how this dynamic process of mediation is accomplished in discursive terms. 'The special qualities of visuals – their iconicity, their indexicality, and especially their syntactic implicitness – makes them very effective tools for framing and articulating ideological messages', Paul Messaris and Linis Abraham observe.[2] Visual imagery, to the extent that it enables journalists to convey meanings that would be difficult to justify in explicit statements, also provides 'a shield of deniability, of a kind that cannot be claimed with verbal persuasion'. Similarly Pierre Bourdieu, in his critique of television news in France, contends that the use of visuals in the news is profoundly ideological. Elaborating this point, he writes:

> The political dangers inherent in the ordinary use of television [news] have to do with the fact that images have the peculiar capacity to produce what literary critics call a *reality effect*. They show things and make people believe in what they show. This power to show is also a power to mobilize. It can give a life to ideas or images but also to groups. The news, the incidents and accidents of everyday life, can be loaded with political or ethnic significance liable to unleash strong, often negative feelings, such as racism, chauvinism or xenophobia. The simple report, the very fact of reporting, of *putting on record* as a reporter, always implies a social construction of reality that can mobilize (or demobilize) individuals or groups.[3]

This 'reality effect' occurs even when the journalist is acting in good faith, Bourdieu maintains. Of particular concern, in his view, are the ways in which journalists allow themselves to be 'guided by their interests (meaning what interests them), presuppositions, categories of perception and evaluation, and unconscious expectations', thereby effectively ensuring that what is offered as a record of reality becomes, in sharp contrast, a creation of it. 'We are getting closer and closer to the point where the social world is primarily described – and in a sense prescribed – by television', in his view.

This chapter's engagement with the visual culture of television news will explore these themes by adopting a historical perspective, and in so doing seek to render problematic the very *naturalness* of its familiar forms, practices and conventions. At stake is the need to deconstruct 'the basic visual vocabulary of television news', as Jenny Tobias suggests, so as to help elucidate how and why certain emergent conventions of visualization gradually became consolidated to the point that they have become recurrent features of newscasts around the globe.[4] In conducting this enquiry, we will limit our focus to pertinent developments in Britain and the United States. Such an approach, it is hoped, will enable a number of historical comparisons to be made by highlighting points of similarity and difference between their respective televisual news cultures. Moreover, it is important to bear in mind that at the time they were initially being established, these two models of broadcasting provided many journalists, editors and producers situated around the world with formative sources of alternative ideas, visual strategies and tactics to use when defining what should count as a proper, authoritative newscast in visual terms. On this basis, it will be shown how notions such as 'impartiality', 'balance' and 'fairness' were encoded as guiding principles for broadcast journalism in these two countries, often in surprisingly different ways.

The pressures of the medium

For reasons that will be made apparent, we begin by noting the passing of veteran broadcast journalist and news anchor, David Brinkley. Brinkley's television career began in 1951 as a correspondent with the *Camel News Caravan*, where he played a key role in introducing television news to US audiences. In October 1956, the *Caravan* was replaced by *The Huntley-Brinkley Report*, which featured co-anchors Brinkley in Washington, DC and Chet Huntley in New York City (Fig. 9.1). As the NBC network's flagship television newscast, it was widely credited for developing a number of innovative features over the years. Following Brinkley's retirement in 1970, the programme was renamed *NBC Nightly News*.

Figure 9.1 *The Huntley-Brinkley Report*

On the occasion of Brinkley's death on 11 June 2003, Tom Brokaw, former anchor of *NBC Nightly News*, recalled: 'David Brinkley was an icon of modern broadcast journalism, a brilliant writer who could say in a few words what the country needed to hear during times of crisis, tragedy and triumph.'[5] Evidently, years earlier, when asked in an interview what he thought his legacy to television news would be, Brinkley had remarked:

> [E]very news program on the air looks essentially as we started it [with *The Huntley-Brinkley Report*]. We more or less set the form for broadcasting news on television that is still used. No one has been able to think of a better way to do it.[6]

In an essay adapted from his memoirs, however, Brinkley's views on the current state of television news could hardly have been more critical. He wrote:

> TV anchors and reporters serve the useful function of delivering the goods, attractively wrapped in the hope of attracting some millions of people to tune in.

In recent decades, I fear, the wrapping has sometimes become too attractive and much television news, in response to economic pressures, competition and perhaps a basic lack of commitment to the integrity and value of the enterprise, has become so trivial and devoid of content as to be little different from entertainment programming. But even at its best, television news is driven less by the ideology of those who deliver it than by the pressures of the medium itself. And as a result, individual journalists, from the anchors to the local news beat reporters, are all constrained in their power by the skepticism of a public that from the beginning saw in television something closer to the tradition of entertainment (movies, theater and the like) than to the tradition of the press.[7]

These words, written by someone who played such a significant part in helping to consolidate the conventions of television journalism, deserve careful attention. Not only do they constitute a warning about the future direction of television news, but also Brinkley makes the crucial point that it is 'the pressures of the medium itself' which are necessarily shaping its ongoing configuration. That is to say, the basic tensions engendered 'from the beginning' between the 'tradition of entertainment' and the 'tradition of the press' continue to inform its development, for better and – clearly in his view – for worse.

Television news, which had first appeared in the United States during the 1930s on several experimental stations, did not get fully underway until after the Second World War. The first regularly scheduled network newscast to adopt the general characteristics familiar to us today was *The CBS-TV News* with Douglas Edwards, which appeared in a 15-minute slot each weekday evening beginning in August 1948 (newscasts would not be lengthened to a half-hour until September 1963). The newscast was sponsored by the car manufacturer Oldsmobile. This first period of television news in the United States established what Tobias has referred to as the convention of the 'head, desk and graphic', which quickly became and has remained a standard news format design: the talking head newsreader sitting at a desk in front of a map, photo or some other graphic.[8]

In February 1948, NBC launched a 10-minute newscast called the *Camel Newsreel Theatre* fronted by John Cameron Swayze and sponsored by Winston-Salem, makers of Camel cigarettes. It featured Swayze reading copy over film; he was seldom actually seen on screen. The following year, NBC reformatted the newscast, extending it to 15 minutes, and renamed it *The Camel News Caravan*. Edward Bliss indicates that 15-minute, half-hour and hour-long programme lengths came to US television from radio, where producers allowed extra time within the newscast slot to sell to advertisers. This signals, from the earliest days of television, a link between the news and its commercial sponsors.[9]

Jenny Tobias suggests that *The Camel News Caravan* was 'arguably the first [television newscast in the US] to take place in a discrete setting and to have a particular visual identity'. Moreover, she contends that visually, the newscast

> combined signifiers for home and office, private and public space. ... The curtains, books and smoking apparatus create a domestic feel, as if one could open a door from one's TV room to Swayze's. The appeal to the domestic may have helped integrate television news into the household.

Yet, the set also had strongly modernist features, with an office desk complete with a nameplate, teletype machine and world map on the wall, suggesting it might be an executive's office or perhaps a war room. The role for Swazye, then, was to mediate the lines between the public world of news events upon which he was reporting for a private, domesticated audience watching in their living-room.[10] Nowhere is this more evident than in the opening and closing taglines, 'Now let's go hopscotching the world for headlines!' before bringing it to a close with his customary 'That's the story, folks. Glad we could get together!'[11]

Most of the editors and reporters who found themselves working in television news had backgrounds in newspapers, wire services or radio news organizations. Such was likewise the case for the producers and production people, although they also tended to be drawn from wire service picture desks, newsreels and picture magazines.[12] The significance of these disparate backgrounds is apparent in the types of debates that emerged regarding how best to present news televisually. In essence, the television newscast represented a blending of the qualities of radio speech with the visual attributes of the newsreel. With little by way of precedent to draw upon, a number of variations on basic newscast formats were tried and tested during these early years.[13]

If the techniques of radio news provided a basis for anchoring the authority of the voiceover, it was the newsreel which supplied a model for the form that television news might take. Aspects of this model included 'the fragmented succession of unrelated "stories", the titles composed in the manner of front page headlines, and the practice of beginning each issue with the major news event of the day, followed by successively less important subject matter'.[14] Newsfilm items tended to be the principal component of the newscast (videotape was first used in network news in 1956), although switches to reporters in other cities were by now a regular feature. The performative role of the 'anchorman' (women were almost always denied this status) was also firmly established by the mid-1950s. One exception to the general rules in play was the early morning *Today* programme on NBC. Its mix of news, features and variety show elements enjoyed wide popular appeal, the latter leading to the inclusion of a charismatic chimpanzee named J. Fred Muggs as a regular member of the presenting team for several years.[15]

By 1954, television had displaced radio in the daily audience figures for usage of each medium, registering just under 3 hours to radio's 2.5 hours according to various surveys.[16] Newscast formats had become relatively conventionalized from one network to the next by this time, although the question of how journalistic notions of 'impartiality' and 'fairness' were to be achieved in practical terms was the subject of considerable dispute. The determined search for ever larger ratings figures, due to the higher sponsorship revenues they could demand, made television news increasingly image-oriented in its drive to attract audiences. An emphasis was routinely placed on staged events, primarily because they were usually packaged by the news promoters behind them (whether governmental or corporate) with the visual needs of television in mind. News of celebrities, speeches by public figures, carnivals and fashion shows made for 'good television', and such coverage was less likely to conflict with sales of advertising time.

Significantly, then, the very features of television news which some critics pointed to as being vulgar, banal or trivial were often the same ones that advertisers believed created an appropriate tone for the content surrounding their messages. Pressure was recurrently brought to bear on the networks to ensure that their viewers, as potential consumers, would not be offended by newscasts presenting the viewpoints of those from outside the limits of pro-business 'respectability'. It was precisely this concern that Edward R. Murrow, the leading news broadcaster on the CBS network, raised when issuing a stark warning in a speech to the Radio Television News Directors Association in 1958. Describing what he considered to be the 'mortal danger' the nation faced, he declared:

> Our history will be what we make it. And if there are any historians about fifty or one hundred years from now, and there should be preserved the kinescopes for one week of all three networks, they will there find recorded in black and white, or colour, evidence of decadence, escapism and insulation from the realities of the world in which we live. ... If this state of affairs continues, we may alter an advertising slogan to read: LOOK NOW, PAY LATER. For surely we shall pay for using this most powerful instrument of communication to insulate the citizenry from the hard and demanding realities which must be faced if we are to survive.[17]

Television, he argued, encourages apathy in its audiences, effectively shielding them from anything disagreeable. He continued:

> We have a built in allergy to unpleasant or disturbing information. Our mass media reflect this. But unless we get up off our fat surpluses and recognize that television in the main is being used to distract, delude, amuse and insulate us, then television and those who finance it, those who look at it and those who work at it, may see a totally different picture too late.[18]

To win the 'decisive battle' against 'ignorance, intolerance and indifference', he believed, would require refashioning television to teach, illuminate and inspire. Otherwise, he concluded, it would be little more than 'wires and lights in a box'.

The discipline of impartiality

Although Britain's first experimental televisual programme was transmitted from Broadcasting House on 22 August 1932, and news had made its appearance on 21 March 1938 (a recording of radio news presented without pictures), newscasts would not be a daily feature on television until 1954. The television service had returned on 7 June 1946, having been closed down during the war years, in part because of fears that enemy bombers would home in on the transmitters. The radio news division prepared a nightly summary of the news to be read on television by an unseen announcer, while a clock-face appeared as the visual component. Newsreels were now manufactured in-house, due to the refusal of the cinema newsreel companies to supply them, and outside broadcasts were also regularly featured.

The BBC, always fearful of the charge that its views were being broadcast in its newscasts, took elaborate care to ensure that it observed a commitment to 'impartiality' as a professional and public duty. Given its responsibilities as a trustee in the national interest, the corporation could not be seen to be expressing a partisan position, especially in matters of public policy. Indeed, anxieties expressed by members of the main political parties that the BBC could ultimately appropriate for itself the status of a forum for national debate to match that of Parliament led, in turn, to the implementation of the '14-day rule' beginning on 10 February 1944 (it would stay in place until 1957). By agreeing (at first informally) not to extend its coverage to issues relevant to either the House of Commons or the House of Lords for 14 days before they were to be debated, the BBC succumbed to pressures which severely compromised its editorial independence. No such restrictions were requested vis-à-vis the newspaper press, nor would their imposition likely to have proven to be successful.

By the early 1950s, with Britain engaged in the war in Korea (filmed coverage of which sparked public interest in the televisual reports), the arrival of competition from the commercial sector in the form of the Independent Television (ITV) network was imminent. BBC officials scrambled to get a daily newscast on the air prior to the launch of the new, commercial rival. Two weeks before the Television Act received the royal assent, the first edition of the BBC's *News and Newsreel* was broadcast on 5 July 1954. While the 7.30 p.m. programme had been heralded as 'a service of the greatest significance in the progress of television in the UK', Margaret Lane, a critic in the corporation's own journal, *The Listener*, was not convinced:

> I suppose the keenest disappointment of the week has been the news service, to which most of us had looked forward, and for which nobody I encountered had a good word. The most it can do in its present stage is to improve our geography, since it does at least offer, in magic lantern style a series of little maps, a pointer and a voice. ... The more I see of television news in fact the more I like my newspaper.[19]

Shortly thereafter, Gerald Barry would comment in his television column in the *Observer* newspaper:

> The sad fact has to be recorded that news on television does not exist. What has been introduced nightly into the TV programmes is a perfunctory little bulletin of news flashes composed of an announcer's voice, a caption and an indifferent still photograph. This may conceivably pass as news, but it does not begin to be television.[20]

By June 1955, the title *News and Newsreel* was dropped in favour of *Television News Bulletin*. The 10 minutes of news was read by an off-screen voice in an 'impersonal, sober and quiet manner', the identity of the (always male) newsreader being kept secret to preserve the institutional authority of the BBC, to the accompaniment of still pictures (as the title suggests, the news was then followed by a newsreel). Only in the final days leading up to the launch of its 'American-style' rival on the new commercial network did this practice change, and then only partially. In the first week of September 1955, the BBC introduced the faces of its newsreaders to the camera, but not their names. The danger of 'personalizing' the news as the voice of an individual, as opposed to that of the corporation, was considered to be serious enough to warrant the preservation of anonymity. This strategy, which had its origins in radio, arguably communicated an enhanced sense of detached impartiality for the newscast, and would last for another 18 months. The policy of anonymous newsreading would continue for BBC radio until 1963.[21]

The Television Act (1954), introduced by Winston Churchill's Conservative government after two and a half years of often acrimonious debate, had set up the Independent Television Authority (ITA).[22] The ITA established, in turn, Independent Television News (ITN) as a specialist subsidiary company in February 1955. Clause 3 of the Act required, among other things, that 'any news given in the programmes (in whatever form) is presented with due accuracy and impartiality'. The imposition of this prohibition within the Act on to the independent programme companies was broadly consistent with the general editorial policy of the BBC. Still, an important difference with respect to how impartiality was to be achieved had been signalled, if not clearly spelt out. Where the BBC generally sought to reaffirm its impartiality over a period of time, ITN would have to demonstrate a 'proper balance' of views within each individual programme.

At 10 p.m. on 22 September 1955, ITN made its debut on the ITV network. The 'newscaster' for that evening, as they were to be called, was Christopher Chataway, a one-time Olympic runner who had been working as a transport officer for a brewery. The other 'personalities' hired by the network included the first female newscaster on British television, Barbara Mandell (a former radio news editor in South Africa), who presented the midday bulletin,[23] and Robin Day, then an unknown barrister with little journalistic experience, who fronted the 7 p.m. bulletin. 'News is human and alive', declared Aidan Crawley, ITN's first editor, 'and we intend to present it in that manner.'[24] This view was reaffirmed by Geoffrey Cox, who assumed the role of editor just months after the launch following the resignation of Crawley over budget disputes with the networking companies. It was Crawley and Cox's shared opinion that 'the power of personality' in presenting the news was a crucial dimension of the effort to attract public attention away from the BBC and on to ITN as a distinctive news source.[25] Here it is also interesting to note that Cox came from a newspaper tradition, namely the London *News Chronicle*, which presumably gave him a different approach to televisual news values than his counterparts at the BBC for whom a background in radio news was the norm.

In contrast with the BBC's anonymous newsreaders, ITN's newscasters were given the freedom to rewrite the news in accordance with their own stylistic preferences as journalists, even to the extent of ending the newscast with a 'lighter' item to raise a smile for the viewer. Cox was well aware, though, that the advantages to be gained by having newscasters who were 'men and women of strong personality' (who also tended to be 'people of strong opinions') had to be qualified in relation to the dictates of the Television Act concerning 'due accuracy and impartiality'. Given that ITN was a subsidiary company of the four principal networking companies, lines of administrative authority were much more diffuse than was the case in the BBC or, for that matter, in the newspaper press. Still, pressure from the networking companies to increase the entertainment value of the newscasts was considerable. As Anthony Smith notes, such pressures, along with the 'invention of a series of technical and creative methods which created a new interest in news which until that time, especially in television, had been little more than a solemn ritual' began to turn journalists into 'national figures'.[26] The visuality of the medium was helping to create reporters with recognizably individual styles of news presentation.[27]

Robin Day, who eventually became one of Britain's most well-known journalists, has credited Cox's editorial standards at the time for securing 'vigorous, thrusting news coverage, responsibly and impartially presented in popular style'.[28] In Day's view, Cox possessed a 'profound belief' in the principles of 'truth and fairness', qualities which meant that under his editorship 'ITN succeeded in combining the challenge and sparkle of Fleet Street with the accuracy and impartiality required by the Television Act'.[29] If this assertion is a somewhat boastful one, it nevertheless reaffirms how, from a journalistic point of view, the tenets of impartiality tend to be rendered as being consistent with professionalism.

This 'discipline of impartiality', with its appeal to the separation of news and opinion, also had implications for ITN's configuration of 'the public' for its newscasts. In its first year, ITN dramatically redefined the extent to which so-called 'ordinary' people could be presented in a televisual news account. Street-corner interviews, or vox pops as they were often called by the newscasters, began to appear on a regular basis. Moreover, at a time when 'class barriers were more marked', Cox recalls that ITN sought to portray the news in 'human terms' through reports which

> brought onto the screen people whose day to day lives had not often in the past been thought worth reflecting on the air. It gave a new meaning to the journalistic concept of the human interest story. In Fleet Street, the term meant stories that were interesting because they were of the unusual, the abnormal, the exceptional. But here the cameras were making fascinating viewing out of ordinary everyday life, bestriding the gap between the classes – and making compulsive television out of it. Whether the story was hard news or not did not seem to matter. It was life, conveyed by the camera with honesty and without condescension, adding interest and humanity to the bulletins in a way unique to this new journalistic medium.[30]

Cox maintains that his sense of ITN's audience at the time was that it was 'largely working class', yet this assumption could not be allowed to 'bias' the network's news agenda. ITN's preferred definitions of 'news values', if not quite as restrictive as those of the BBC, still ensured that a potential news source's 'credibility' or 'authoritativeness' would be hierarchically determined in relation to class (as well as with regard to factors such as gender and ethnicity).

By 1956, the BBC had elected to follow ITN's lead. In seeking to refashion its televisual newscasts to meet the new 'personalized' standards of presentation audiences were coming to expect, the corporation began to identify its newsreaders by name. It also emulated ITN by allowing them to use teleprompters in order to overcome their reliance on written scripts. Further technological improvements, most notably in the quality of film processing, similarly improved the visual representation of authenticity. That said, the question of whether or not to use dubbed or even artificial sound to accompany otherwise silent film reports posed a particularly difficult problem for journalists anxious to avoid potential criticisms about their claim to impartiality. Much debate also ensued over what circumstances justified imitating ITN's more informal style of presentation, particularly with regard to the use of colloquial language, to enhance the newscast's popular appeal (previously BBC news writers had been told to adopt a mode of address appropriate for readers of the 'quality' press). ITN had also shown how the new light-weight 16mm film camera technology could be exploited to advantage 'in the field' for more visually compelling images (complete with 'natural sound') than those provided by the newsreel companies with their bulky 35mm equipment. Indeed, through this commitment to 'bringing to life' news stories in a dramatic way, as well as its more aggressive approach to pursuing 'scoops' (exclusives) and 'beats' (first disclosures), ITN was stealing the march on the BBC with respect to attracting a greater interest in news among viewers.

Researching media history

It is striking to note the extent to which certain familiar accounts of television news's rise to prominence over the years make each sequential development sound almost inevitable, as if they were unfolding in relation to a proscribed imperative. One would be forgiven for thinking that the history of the medium is best understood in technological terms, where improvements in technical capacity will necessarily engender a better quality of news reporting. By this logic, television news today is extraordinarily impressive. Its visual culture revolves around an array of innovations virtually unimaginable in the 1950s, not least with respect to the advent of digital technologies – the perceived virtues of which are celebrated in new rhetorics of liveness, immediacy or

Figure 9.2 TV satellite vans relay the news in an instant

authenticity – which make reports of breaking news possible from anywhere in the world in a near-instant (Fig. 9.2).

As we have endeavoured to show in this chapter, however, in seeking to identify the subtle, often contradictory imperatives informing the gradual consolidation of reportorial conventions, a historical perspective is invaluable. Moreover, such a perspective calls into question the easy assumption that technological progress consistently translates into more advanced forms of journalism. The US news broadcaster Robert MacNeil, in his account of the early days of the medium, recalls the words of the NBC network's then chairperson, Walter Scott, on what happens when 'the true reporting function' is obscured by technical innovations in delivering visuals for their own sake. 'Because television *is* a visual medium,' Scott stated, 'it may scant the background and significance of events to focus on the outward appearance – the comings and goings of statesmen instead of the issue that confronts them'.[31] MacNeil concurs, reminding us that 'from its inception television news has been criticized for a tendency to let pictures dictate the story', before making a sharper criticism:

> Television newsmen cannot be blamed for wanting to put visual material on a visual medium, but when preoccupation with visual effects overrides news judgement, it encourages emphasis on action rather than on significance and the playing up of trivial or exciting occurrences simply because they can be covered by cameras [Fig. 9.3].[32]

Figure 9.3 Camera crews compete for the best pictures

In other words, news reportage's capacity to deliver images – no matter how compelling or heart-rending – will represent a hollow achievement without adequate context, explanation or interpretation. Even worse, this preoccupation may actually be harmful to the extent that it impedes public understanding of the events in question.

Much work remains to be done where investigations of the visual culture of television news is concerned, of course.[33] In addition to the conceptual difficulties highlighted above, there is also a range of practical matters that invite careful consideration. Media historians must strive to be ever-more self-reflexive, not least with respect to the normative criteria informing their chosen strategies when gathering source material and interpreting evidence. Equally pertinent is the recognition that the journalistic processes under scrutiny are ephemeral, and thereby more than likely to prove elusive in conceptual and methodological terms. Indeed, it is their very normality, that is, the extent to which they were simply taken for granted as 'common sense' in the newsroom and in the field, that makes efforts to de-normalize them so challenging. To the extent that it is possible to discern the contours of the imperatives giving shape to visual culture, especially with respect to the economic, political and cultural dynamics which imbue its logics, it is likely that they will be more apparent in retrospect than they were at the time. In other words, from the vantage point of today, it is vitally important to appreciate the socially contingent, frequently contested nature of their lived negotiation – and thereby resist any sense of inevitability in our discoveries.

Notes

1   See also Stuart Allan, *News Culture*, 2nd ed. (Maidenhead: Open University Press, 2004). Please note that this chapter draws on material from this source.
2   Paul Messaris and Linis Abraham, 'The Role of Images in Framing News Stories', in *Framing Public Life: Perspectives on Media and our Understanding of the Social World*, ed. Steven D. Reece, Oscar H. Gandy and August E. Grant (Mahwah, NJ: Lawrence Erlbaum, 2001), 220.
3   Pierre Bourdieu, 'Television', *European Review* 9, no. 3 (2001): 248–9. Emphasis in the original.
4   Jenny Tobias, 'Truth to Materials: Modernism and US Television News Design since 1940', *Journal of Design History* 18, no. 2 (2005): 179–90.
5   CNN.com, 'Pioneer newsman David Brinkley dies at 82', 12 June 2003, www.cnn.com/2003/SHOWBIZ/TV/06/12/obit.brinkley/index.html.
6   Clayland H. Waite, 'David Brinkley', *The Museum of Broadcast Communications*, www.museum.tv/archives.
7   David Brinkley, 'On being an anchorman', *The New York Times*, 14 June 2003.
8   Tobias, 'Truth to Materials', 180.
9   Edward Bliss, *Now the News: The Story of Broadcast Journalism* (New York: Columbia University Press, 1991), 223.
10 Tobias, 'Truth to Materials', 182.
11 Erik Barnouw, *Tube of Plenty*, 2nd ed. (New York: Oxford University Press, 1990), 102–3.
12 T. Nielsen, 'A History of Network Television News', in *American Broadcasting: A Source Book*, ed. Lawrence W. Lichty and Malachi C. Topping (New York: Hastings House, 1975).
13 See Kevin G. Barnhurst and Catherine A. Steele, 'Image Bite News: The Visual Coverage of

Elections on U.S. Television, 1968–1992', *Howard International Journal of Press/Politics* 2, no. 1 (1997): 40–58; Michael Schudson, 'The Politics of the Narrative Form: The Emergence of News Conventions in Print and Television', *Daedalus* 111 (1982): 97–112; Tobias, 'Truth to Materials', 179–90.

14 Fielding cited in Brian Winston, 'The CBC Evening News, 7 April 1949: Creating an Ineffable Television Form', in *Getting the Message: News, Truth and Power*, ed. John Eldridge (London: Routledge, 1993), 184.

15 Barnouw, *Tube of Plenty*, 147–8.

16 David C. Bianculli, *Teleliteracy: Taking Television Seriously* (New York: Touchstone, 1992), 58.

17 Edward R. Murrow, 'Wires and Lights in a Box', in *Documents of American Broadcasting*, ed. Frank J. Kahn (Englewood Cliffs, NJ: Prentice-Hall, 1978), 253–4. Originally presented as a speech to the Radio and Television News Directors Association, Chicago, IL, 15 October 1958.

18 Ibid., 260.

19 Cited in Geoffrey Cox, *Pioneering Television News* (London: John Libbey, 1995), 38.

20 Cited in Anthony Davis, *Television: Here is the News* (London: Severn, 1976), 13.

21 See Grace W. Goldie, *Facing the Nation: Television and Politics, 1936–1976* (London: Bodley, 1977); Phillip Schlesinger, *Putting 'Reality' Together: BBC News* (London: Methuen, 1987), 37. See also Asa Briggs, *The History of Broadcasting in the United Kingdom*, vols 1–5, *The Birth of Broadcasting* (London: Oxford University Press, 1961–95); Valeria Camporesi, 'The BBC and American Broadcasting, 1922–55', *Media, Culture and Society* 16, no. 4 (1994): 625–39.

22 John Reith, 'Speech in Debate on Commercial Television', in *British Broadcasting*, ed. Anthony Smith (Newton Abbot, UK: David & Charles, 1974).

23 This bulletin featured a painted screen backdrop behind Mandell picturing a kitchen with a sink full of dirty dishes, thus visually reaffirming her position within the private sphere and undermining her credibility as a neutral voice of authority; see Janet Thumim,'"Mrs Knight *Must be* Balanced": Methodological Problems in Researching Early British Television', in *News, Gender and Power*, ed. Cynthia Carter, Gill Branston and Stuart Allan (London: Routledge, 1998), 97.

24 Cited in Anthony Hayward, 'Obituary: Barbara Mandell', *The Independent*, 5 September 1998; see also Aidan Crawley, *Leap Before You Look* (London: Collins, 1988).

25 See also Burton Paulu, *British Broadcasting in Transition* (London: Macmillan, 1961); Bernard Sendell, *Independent Television in Britain*, vol. 1 (London: Macmillan, 1982).

26 Anthony Smith, *The Shadow in the Cave: The Broadcaster, the Audience and the State* (London: Allen & Unwin, 1973), 77.

27 Ibid., 78.

28 Robin Day, foreword to *Pioneering Television News*, ed. Geoffrey Cox (London: John Libbey, 1995), viii; see also Robin Day, *Grand Inquisitor* (London: Severn, 1989).

29 Day, foreword to *Pioneering Television News*, ix.

30 Cox, *Pioneering Television News*, 57.

31 Walter Scott cited in Robert MacNeil, *The People Machine: The Influence of Television on American Politics* (New York: Harper & Row, 1968), 35.

32 Robert MacNeil, *The People Machine: The Influence of Television on American Politics* (New York: Harper & Row, 1968), 35.

33 See Richard Howells, *Visual Culture* (Cambridge: Polity, 2003); Barnhurst and Steele,

'Image Bite News'; Barbie Zelizer, 'Going Beyond Disciplinary Boundaries in Future Journalism Research', in *Global Journalism Research*, ed. Martin Löffelholz and David Weaver (Oxford: Blackwell, 2008), 253–66.

10 'What planet are we on?' Television drama's relationships with social reality

by Máire Messenger Davies

In his chapter on television in *Visual Culture*, Richard Howells raises the question of television realism and asks: 'If aliens from the planet Tharg were to be monitoring the earth by watching our television programmes, how realistic an impression of our lives would they receive?'[1] This intriguing prospect is exactly paralleled in the film *Galaxy Quest* (1999).[2] In this film, a group of aliens come down to earth and kidnap the actors from a TV space-opera, pretty obviously based on the TV science-fiction series, *Star Trek* (of which more later). The aliens believe that the Galaxy Quest crew are genuine space-warriors and much humour ensues when the motley company, led by their drunken leading man (Tim Allen) and an ego-driven English Shakespearian actor (Alan Rickman), find themselves actually fighting space-monsters. The denouement of the film brings 'reality' and fantasy together and exuberantly demonstrates that 'truth', if not 'reality', is compatible with make-believe; it illustrates Schiller's remark that 'the fairytales of my childhood have a deeper meaning than the truths taught by life'.[3]

Howells' description of some of the weird TV material the Thargians could encounter suggests that using television as a source of evidence about anything could lead to some major misunderstandings. As a 'dumbed-down' medium which, in Neil Postman's noted phrase, is 'amusing us to death' and distracting us from more important issues in the real world beyond, the entire medium of television, not merely its dramatic storytelling, could simply be dismissed as suspect.[4] This, however, would be a mistake. Even if – indeed, especially if – the medium is at odds with supposedly more objective sources of information about society, such as printed texts, official documents, eye-witness reports and archaeological finds, it should still be critically examined, if only as a source of these supposedly 'dumbed-down' misperceptions in the audience. But more to the point, dramatic fiction is historically the most frequently recurring form of representation in the major mass medium of the second half of the twentieth century and, despite the advent of the Internet, in the twenty-first. Since television became widespread and popular in the United States in the 1950s and in the UK during the 1960s, TV drama has come to comprise many different genres – soap opera, westerns, science-fiction, fantasy, costume drama, literary adaptations, social realism issue dramas, docudrama, single plays, mini-series, situation comedies and so on – and there are thousands of different texts within each of these different genres. Despite a recent trend towards 'reality television', fictional dramatic storytelling remains the most durable form of TV representation. Soap operas such as *Coronation Street* and *EastEnders* remain high in British audience ratings and TV drama continues to attract the largest audiences. Thus, scholars of human society – whether historians, sociologists, psychologists or critical textual analysts – must certainly take this huge body of evidence of human activity seriously. This chapter refers primarily to British and American television, and will focus on a small number of illustrative examples.[5]

The artifice of drama is the first fact that must be borne in mind when approaching the body of evidence supplied by TV drama. A study of television drama I carried out with 1300 7–12 year olds showed that children overwhelmingly chose as their definition of drama 'a story with actors or people dressed up' and, given the option of creating their own definitions, references to acting and pretence occurred repeatedly. Responses included:

> 'Drama is pretending to be someone else.' Boy, 8, Cardiff.
> 'Drama is something that people act out.' Girl, 8, Cardiff.
> 'Drama is a programme where people get dressed up to act a play.' Girl, 12, Co. Durham.
> 'Drama is people acting.' Girl, 9, Outer London.
> 'Drama is a real event that is acted out again in a series like *London's Burning* and *EastEnders*.' Boy, 12, Milton Keynes.[6]

Even primary school children recognize that drama is 'pretend', and as such, that it needs to be judged by criteria which include a recognition of its artifice – that its sense of verisimilitude is created by illusion and that these illusions are derived from the codes and conventions of the medium, which can be identified and named. This ability to identify artifice in evaluating the differences between 'real' and 'pretend' has been described as 'modality awareness'. Although television drama can be 'about real life' it never actually is real life – and not being real life was one of its charms, as a child in Cardiff pointed out: 'It lets your mind escape.'[7]

As these children were aware, TV drama consists of made-up stories, sometimes based on real events, constructed by writers, performed by actors, staged by producers, directors and various technical crew members. Further artificial elements include the fact that the performance is recorded electronically (although, in the case of sitcom, sometimes in front of a live audience) and distributed in a variety of artificial formats, including broadcasting to air, DVDs and online streaming. In these ways, it differs fundamentally from live theatrical performance, with which it otherwise shares many common elements. Awareness of television's dramatic techniques and artistic constructions, and how they themselves change over time, partly as a function of political, economic, cultural and technological developments, needs also to be built into any scholarly examination of television drama as evidence.

What does television drama look like? Visually, the formal features of television drama include:

1 *Framing*: all action has to take place within the relatively small dimensions of a domestic television screen; hence dramatic spectacle works less well than domestic intimacy, with frequent close-ups of faces. This gives rise to the dominant modes of television drama being domestic and intimate (soap opera

being the obvious example). Even expensive and relatively spectacular action series like *24* or *Lost* incorporate detailed storylines of intimate relationships, love affairs, marriages, children, friendships and, especially, relatively large amounts of dialogue compared to feature film.

2 *Human action*: the movement of characters through space.
3 *Camera movement*: panning, zooming, cutting (editing).
4 *Camera effects*: dissolves, fades, ripples, multiple screens.
5 *Casting*: the embodiment of fictional characters by particular kinds of people. Casting has been used as evidence of racial prejudice and sexual stereotyping by some scholars. Such scholars use the technique of content analysis to count, for example, the number of times women or ethnic minorities appear, or have leading or authoritative roles. George Gerbner and his colleagues at the University of Pennsylvania's Annenberg School used content analyses of American primetime drama, carried out from the 1960s to the 1990s, to show, for example, that men outnumbered women three to one. They also found that older people, younger people, blacks and Hispanics were under-represented in television drama, compared to the general population; and that crime was ten times as frequent in the 'television world' as it was in the real world. In TV drama, certain groups were also more likely than others to be victims (women and ethnic minorities) and violence was more frequently carried out by male authority figures, which, argued the researchers, tended to legitimize it. Content analysis of the kind carried out at Annenberg is one of the most frequently used and systematic ways in which television drama has been used as evidence for social attitudes.[8]
6 *Design (animate)*: costume, hair and make-up.
7 *Design (inanimate)*: scenery, furniture, props, lighting.
8 *Mise-en-scène*: the general visual set-up which incorporates framing and other design elements.
9 *Special effects*: including computerized effects, usually added in post-production.
10 *All mediated by the visual perceptions of viewers*: Planet Thargians will see things differently from earth-based TV viewers; small children see things differently from older ones.

As Gavriel Salomon points out, television is a multiple symbol system, using a variety of semantic codes – both visual and aural – simultaneously, all of which need to be decoded together by viewers.[9] Until the invention of videotape, and later DVD, it was not possible for audiences to review what they had seen and heard or for researchers to examine texts repeatedly. Evidence had to be drawn from people's memories or from anecdotal accounts of first-hand experiences, for instance Raymond Williams' famous account of experiencing 'flow' as he watched American television.[10] Alternatively, researchers had to use correlational research, like that of the Gerbner team at Annenberg, positing a relationship between self-reported 'heavy viewing' (more than four hours a night) with a 'mean world' view, arising from all the violence that such people were assumed to have watched in primetime drama.

Now that material is generally available for review, the visuals of television can be analysed in more depth – for instance, semiotically – applied by practitioner-theorists such as John Ellis and outlined for the purpose of media teaching in the way that texts like James Monaco's classic *How to Read a Film* have long done for the cinema.[11] Some of the most useful explanations of how television works visually have been written by practitioners, for example George Millerson's *The Technique of Television Production* (1990, now in its twelfth edition).[12] Practitioners' concerns are necessarily more with technical competence and aesthetic effectiveness than underlying symbolism or ideological readings, but such considerations will not be completely absent from their thinking, as interviews with production staff can reveal. More recently, aesthetic concerns, such as lighting, sound quality and design have begun to be recognized and analysed by TV scholars such as Karen Lury.[13] Thus, television drama as visual evidence consists not only of analysis of the texts themselves, but also of production criteria, as expressed by producers as well as researchers, and analysis of audience responses.

Bearing in mind the proviso that dramatic stories are make-believe and electronically packaged, and employing a variety of visual codes and conventions which are constructed and artificial, we can derive some historical and social evidence for some human attributes from TV drama. Television dramas such as *The Cosby Show* have frequently been analysed for the social impact of their actions and dialogue. In such cases, the main source of evidence is the events of the narrative and the characters represented in it. But there are other sources of evidence that are not exclusively narrative based. In contemporary dramas (set in the period in which they were produced), TV can show us what people (or at least actors) looked and sounded like, what they wore and what was fashionable at the time. The more popular the show, the more likely it is that these representations will be designed to be recognizable and identifiable to the largest possible group of people in the population – and hence reasonably representative of their tastes. This will not necessarily be true of more esoteric art forms, but television is a mass medium, with even small audiences representative of all social groups.[14]

Television drama can also suggest evidence of what people thought, what was ideologically acceptable, or on the other hand, controversial. 'Pushing the boundaries' is a popular concept among TV drama producers, for instance in the case of the Channel 4 drama *Queer as Folk*. But the popularity of this show suggests that the boundaries may already have been 'pushed' for many people in the audience; the show's representation of homosexuality was likely to have been 'pushing' at an open door. Thus, the televised text of *Queer as Folk* can best be seen as evidence of increasing public acceptance of homosexuality in conjunction with contextual, non-visual sources of evidence such as its audience ratings and contemporary legislative developments.

Contemporary drama also provides evidence of current styles in fashion, music, demotic speech, household goods, modes of transport and sexual attractiveness. The more up-to-the-minute these are, the more likely they are to appear dated and laughable later on. The BBC's hit drama series, *Life on Mars* (2007) and *Ashes to Ashes* (2008), in which police officers from the 2000s find themselves unexpectedly relocated to police stations in the 1970s and 1980s, rely for much of their entertainment value on the memories of older members of their audiences for the Ford Escorts, the frizzy perms and the David Bowie soundtracks of those periods. Younger audience members may find these allusions puzzling – and historical.

Examining television drama texts can also yield evidence of the changing nature of the medium of TV itself, for instance the change from live to recorded drama and the effects of this on modes of performance and production.[15] The problem of the survival and archiving of early dramas to review is a vexed one and should make us cautious about sociological and psychological claims about the social impact of television texts, which cannot be verified by reference to the texts themselves. There is a problem in carrying out further detailed textual analysis of many TV shows, particularly those from periods before video and DVD, because copies do not survive. Archive research is becoming increasingly important in television studies.[16] For example, the Arts and Humanities Research Council (AHRC) and the BBC set up an archive pilot study in 2006 and the AHRC is also funding a year-long training programme for postgraduate students in the use of moving image archives in research.[17]

Some holders of TV archives, such as the BBC, are belatedly realizing what treasures they have and are offering increasing numbers of programmes for sale. Unfortunately, only prestigious productions that will generate healthy sales are likely to be released rather than the more historically interesting but less commercially profitable ones. This is often true of programmes aimed at minority audiences such as children. *Watch with Mother*, the BBC's 1950s preschool series, is very revealing of the assumptions made about small children's lives at that time, especially the assumption that 'mother' would always be there.[18] The BBC children's drama series, *Grange Hill*, cancelled in 2008 after 30 years, mirrored the changing concerns of several generations of British schoolchildren from the late 1970s onwards, as they grew up from childish first-years to quasi-adult sixth-formers. Often controversial, the programme was greatly loved by children, particularly in the first decade of its existence. It 'dignified the lives of children by connecting their experiences, interests and commitments to wider social currents … conflicts around "race" and around different conceptions of the feminine'.[19] As a BBC programme, it is likely that *Grange Hill* will at least survive in some form. A serious problem for researching the social effects of television representations is that private corporations, which frequently change ownership, do not keep systematic archives in the way that the BBC does and even very expensive and therefore prestigious productions can vanish almost completely.[20]

Social realism: textual, ideological analysis

Does TV drama show society as it really is? *The Cosby Show*, aired in the United States from 1984 to 1992, provides a useful case for the textual and ideological analysis of a TV drama based on genre, casting and narrative events during the period it was broadcast. It can also demonstrate the nature of debate between scholars and other commentators over the historical and political relevance of a television text. In *Enlightened Racism* (1992), published at the end of *The Cosby Show*'s run, Sut Jhally and Justin Lewis point out the gap between real black people's lives in the United States and that of the comfortably off Huxtables, the family in the show.[21] They also raise the influence of 'effects': did Cosby make black people's lives better? One answer has to be that it depends on the scale of the effects you are measuring. Obviously *The Cosby Show*'s coincidence with the Los Angeles riots in 1992, as noted by Jhally and Lewis, suggests that politically and economically, race was still a major problem in the United States in the 1990s, despite Bill Cosby's efforts to generate a more widespread, enlightened attitude to the social position of American blacks. However, to cite a smaller-scale, domestic effect, black people I knew when in the United States loved the show. One friend, who did not own a TV, made a point of visiting me every Thursday so she could watch. She said it was also very popular in her homeland, Suriname, where she and her friends felt its effect of enhancing their sense of identification with these attractive and amusing people, realistically portrayed in terms of such themes as marital one-upmanship or teenage frictions. It did not insult them by portraying black people as only marginal, ill educated or deprived (although such characters did occasionally feature in the show) or – as was the case in much primetime drama – by leaving out people like them altogether.

At the same time, the programme was entertaining and funny, as intended by its writers and performers (Cosby was, after all, a stand-up comedian). Here the issue of genre is important in the quest for evidence of social and political media effects. An archaeologist unearthing evidence of Mayan civilization would not necessarily expect to find evidence of priestly rituals in pieces of kitchen pottery. *The Cosby Show* was a domestic sitcom, with clearly signposted elements of artifice such as soundtrack laughter, not an issue-based drama, nor a documentary. As a sitcom, it belonged to a typical and well-established primetime format in American TV from *I Love Lucy* (1951) onwards. To the extent that the familiar, conventionalized TV landscape of primetime comedy drama was successfully colonized over a period of eight years by a black family and their friends, it could be argued that the show represented an important step in the normalization, as distinct from the 'othering', of blackness in American culture.

Similar comedy-based 'normalizations' have occurred with other marginalized groups on American television: working-class, non-beautiful people (*All in the Family, Roseanne*); working women (*Mary Tyler Moore, Rhoda, Designing Women*); single mothers (*Murphy Brown, Grace under Fire, Friends*); elderly women (*The Golden Girls*); gays (*Will and Grace, Ellen*) and Hispanics (*Ugly Betty*). Certainly, a universal public acknowledgement on primetime TV that fat, working-class people have rights and emotions, and that elderly women can be witty and attractive, is not the ultimate pinnacle of social progress. Such normalization may not have been as powerful as, for instance, direct political action to reduce economic inequality and to safeguard women's pensions. But, if popular culture is assumed to have any influence at all, it can be seen as a necessary, if not sufficient, accompaniment to the advancement of social rights.

Researchers must also imagine the control conditions: would the position of blacks or working-class women be better if there were no positive televisual representations of them? Were their positions better before television was invented? Are they better in societies without access to television? The belief that television has powerful influence on social behaviour is open to question if a programme like *The Cosby Show* had no obvious socially transforming effect. There is, in fact, little compelling evidence that TV watching does have such powerful effects, at least not in directly measurable and publicly obvious ways.[22] One could agree with the conclusions of Patrick Barwise and Andrew Ehrenberg that TV-watching is something you do when you have not got anything more pressing to do, and that its main effect has been to cause people to sit down for several hours every evening. In the case of *The Cosby Show*, Jhally and Lewis's linkage of the programme with major national events and my own observation of domestic consumption suggest otherwise. But no causal link can be empirically demonstrated between the show and any of these outside events; all we can do is suggest possible relationships. This in itself is evidence of the show's impact.

Non-realistic drama: science fiction – *Star Trek*: production studies

When it is acknowledged that no television drama can be viewed as an unproblematically realistic 'window on the world', it follows that social realism, the representation of contemporary people in recognizably realistic settings, becomes less privileged as a source of evidence. Non-realistic dramas such as science fiction and fantasy can also be seen as potential sources of evidence about human societies, when evaluated carefully. The science-fiction series *Star Trek* is one of the most exhaustively analysed of TV drama texts. The five *Star Trek* series, comprising over 700 episodes, aired over a time span of nearly 40 years, from 1966 to 2005: *Star Trek*, The Original Series (TOS): 1966–9 and shown in syndication ever since (Fig. 10.1); *Star Trek: The Next Generation* (TNG) 1987–94; *Star Trek: Deep Space Nine* (DS9) 1993–9; *Star Trek Voyager* (VOY) 1995–2001 – all still in syndication; 2001–2005: *Enterprise* (ENT) – cancelled after four seasons, but still shown in syndication. There were also ten feature films, the most recent being *Nemesis* (2002), and Star Trek (2009)

Figure 10.1 'Mr Spock' from the original *Star Trek* television series

Although *Star Trek* is a self-proclaimed non-realistic science-fiction series, set in the future and peopled with unlikely looking aliens and robots, much of the commentary on it has been textual analysis of its real-world relationship with political ideology, for instance TOS's representations of the Vietnam War and TNG's representations of Cold War diplomacy.[23] As well as scholarly commentary, there is also a vast metaverse of fan commentary, starting with printed fanzines in the 1960s, and later blogs, websites, fanfiction and official *Star Trek* publications of various kinds; this material can also be analysed for evidence of the show's impact and its relationships with broader social trends, although the fan material is far from being a representative audience study. Nevertheless, all this is evidence of how a non-realist television drama can express and stimulate contemporary political and ideological concerns.

Much of the academic and audience commentary focuses on the show's representations of race; gender and 'otherness' generally – something which science fiction as a genre is particularly well equipped to do.[24] By introducing alien forms, including robots and androids, questions of what it means to be 'human' and consequent issues of rights and oppression can be dramatically examined. Because the 700-plus episodes extend over a period of 40 years, *Star Trek* is an unusually comprehensive set of texts for examining how changing attitudes to politics, race, gender and human rights have been dramatically represented to a primarily American (but also international) audience in the years between 1966 and 2005. Its effects are almost incalculable, though it is observable that analogies and jargon from the series – 'boldly go', 'life but not as we know it', 'beam me up' and so on – have entered the culture and have even stimulated people to choose careers in space research.[25]

Roberta Pearson and I have conducted extensive research on the show at Paramount Studios in Hollywood, one aspect of which focused on its status as a television text. Our research included interviews with actors, executive producers, line producers, publicists, writers, directors, designers, set-dressers, make-up artists, set builders, special effects workers and post-production workers. Interviews must be regarded with caution as sources of reliable evidence because people generally present themselves in a way that they hope will make them look as effective as possible, and suppress information that does not. They also need to be discreet in talking about other people with whom they are working. On the other hand, what we learned from these people, we believe, represents a useful body of evidence as to how television drama at that time operated in representing stories and issues and specifically how the genre of science fiction can do this. The interviews also provide evidence of television workers' assumptions about the value of their own work and its possible impact on audiences. Such oral evidence can be interpreted differently at different times; indeed, the evidence here needs to be evaluated in conjunction with copies of the programme itself. The designers' comments about the visuals of television refer partly to the television content – aspects of narrative of concern to textual analysts, such as social acceptability and impact. But they are of particular interest as evidence about the way television worked as a medium at this particular historical moment.[26]

The makers of the 'look': interviews with designers

Herman Zimmerman, the chief designer on the TV franchise, was working on the feature film, *Star Trek: Nemesis* (2002) when we interviewed him. He stressed the fundamental constraint of framing, comparing the TV screen to the movie screen, and illustrating how the shape and size of the screen can determine content:

> I just barely finished *Enterprise* [the TV show] when we started preproduction on the feature [*Nemesis*]. Features are camera images that are as much as 70 feet wide and 30 feet high, as opposed to 19 inch diagonal television screen. So a television is much more forgiving of detail. Most necessarily it's a close-up medium. Both mediums are, but you're going to see a lot more obviously on a cinemascope screen, than you are on your screen at home. So it's partly a sharpening of the quality to really get as much of the money on the screen as you can.[27]

Even in a science-fiction show set in the future (the twenty-fourth century) issues of realism are inescapable. When asked, 'How do you design the future?', Zimmerman laughed and said:

> Well, you design the future by looking at the present. You design the future by telling yourself what is probably not going to change. Even a thousand years from now, we'll still have two eyes, two ears, hands and feet. Have to sleep, have to eat. Tables and chairs. The proportions of those things are going to remain the same, and have remained the same since recorded history, so you can hang your hat on things that are staple. Then you look to the writer for what the basic clue is.[28]

Zimmerman's comments also provide evidence of work practices in television, emphasizing the medium's team-based nature. In addition to 'looking to the writer' he 'found inspiration' in the work of Michael Westmore, chief make-up artist:

> A prosthetic for an alien face gives me an idea of what that alien race would look like. And maybe some of those elements end up being line elements in a design or colour elements in a design or texture.

Costume designer Robert Blackman's costumes for a race called the Cardassians in the later episodes of TNG gave Zimmerman 'the inspiration of doing the whole design for the texture of *Deep Space Nine*'. Zimmerman's remarks are revealing of individual design decisions: 'inspiration'. But they also reinforced our ethnographic impressions from visiting the *Star Trek* production base at Paramount Studios, of television's essentially collaborative nature, and how the creative agency of individuals has to function in the production of modern audiovisual media. Clearly from Zimmerman's remarks, some have a considerable degree of autonomy, but even at a very senior level, such as his, this was always within a collaborative framework.

'Designing the future' and its dependence on the present was reinforced by our interview with one of Zimmerman's sources of inspiration, the costume designer Robert Blackman, who (like several other members of the *Star Trek* design team) had been trained as a theatre designer and had worked on many classical stage productions. He had been with the franchise for 13 years when we spoke to him. He told us that when he started the job, he was asked what the job was like and he answered, 'Well, it's like doing Elizabethan doublets and dance wear.' Blackman was entertaining on the subject of having to design streamlined, futuristic costumes without pockets or proper fastenings, to which he objected on the grounds of both realism and functionalism:

> One of the things where I think my strength is, is I keep saying, no, it's only 150 years from now. So not 150 years from now, it's *only* 150 years from now. So if we think that in 1420 the button was invented, and they were using a form of button and loop then, and it's now 2001 – the zipper's going to be gone in our space suits? No. Our space suits should have 13 zippers, which is what they have.... The zipper came in about 1920, it was a World War One invention, same as the wrist watch, which is a World War One invention. So we should keep those. We should not take them away.[29]

Blackman prided himself on being the person who reminded the team generally about historical continuity: 'One of the things that I do best, which I think would be missed, is my insistence that we remember we are closer to 1921 than we are to 2150.' He also had some entertaining comments on the changes in social acceptability of sexual and physical representations and his discussions with senior executives about this.

> [in the decontamination scene in the pilot] they [senior producers] were all very nervous because we got, not specific shapes, but you saw generalized shapes, you do see the shape of her nipple through the T-shirt. If you turn on television in America at 8 o'clock, you see the shape of a nipple through anybody's T-shirt – it's just everywhere.... but they [the producers] just got fidgety, they have children, they don't want to look like they're doing sexual pandering that their 8 year olds or their 10 year olds are going to watch. I understand it completely. On the other hand, have you seen a music video anywhere? Have you seen Britney Spears anywhere? Give me a break. We're two generations behind the current generation, and we just need to step forward.[30]

Star Trek, over its 40 years of existence, is the source of a great deal of visual evidence about changes in female fashions and in what is seen by contemporary style-leaders as glamorous. The miniskirted, backcombed, Day-Glo outfits of the 1960s actresses in the original series have been replaced by the functional (but still skin-tight) catsuits of the 'Vulcan' T'Pol (Jolene Blalock) in *Enterprise* in the 2000s. In fact, Blackman expressed concern at the increasing pressure on actresses to be thin, another change from standards of glamour in the 1960s. The show's representation of gender, including its casting of exclusively beautiful actresses, has been widely commented on; its designers are, and have always been, entirely aware of the calculatedness of 'the babe factor'. However, for designers, and hence for audiences both male and female, it is important that the glamour seen on the screen should not be too far away from contemporary norms of what female attractiveness is currently assumed to be – and these norms are what is valuable as historical evidence. They can certainly change over time, as Jim Mees, the set-dresser noticed:

My joke about [the sci-fi series] *Babylon 5*. Why did we go hundreds of years in the future and have such big hair? They all look like they came from the beauty parlour planet. There's a real balance between hokey and believability. No one would have believed us if what we had done was just gone back on the old show and made it [the new show] all purple and green.[31]

Evidence of technical changes in television: special and visual effects

In no area of production has there been as much change as in the creation of special effects. Dan Curry, the visual effects producer, who had been with the show for 14 years, had gone from making 'miniatures' – carefully hand-crafted paper models – to managing a completely computerized system. He outlined and also commented on this historical progression for us, suggesting that ingenuity in design was now being lost because of computers:

> There's a whole range of visual effects: starting with what you see on the little monitors. We call those bullions, because in the old days they would have been composited on an optical printer. A lot of the terminology is based on op science and technology. We're even calling them opticals which are not opticals by any sense of the word any more. . . . Computers have changed things radically, both for good and bad. The good news is that we used to spend sometimes 60–80 hours a week keeping up with the miniatures. . . .
>
> I'll tell you what they lack. The computer has taken away the wonderful sense of playful alchemy that we would have. Like, we need a solar explosion. OK, well, if we take baking soda and throw it out of this thing and have it hit on a bowling ball, and then we'll turn it around and squeeze the ball, then we can make it look like solar ejections are hitting the force field on the Enterprise. So that will never ever happen again . . . you had to make it up as you went along, and you needed a sense of seeing whatever materials were around, seeing the potential in those things, apart from their original intent. For example, I can't go to a hardware store and go to the plumbing supply without seeing a space ship part.[32]

Curry also had things to say about the social impact of the technological changes in television creativity, and here began to invoke theories of childhood which are echoed by other commentators in the increasing body of recent work on 'new' media and children, some of it highly critical, for instance Sue Palmer's *Toxic Childhood*.[33]

> There are a couple of things that are detrimental about the world of computers, and that is that they have become ubiquitous in the society of young people; with the games and stuff like that, their imaginations are channeled by the nerds who designed these games. . . . And so, they wish they were ninjas, like, for every hour my son spent playing video karate, he could have had a real black belt.[34]

Curry's views extended to pedagogy, and these are relevant to those of us who teach production students:

> And the fact that they don't build physical models, they don't take photographs, especially now with the digital photography coming up, they don't think about exposure or zones, and they don't play with toys as much as they used to. And I think toys are really critical to spurring the imagination, because the child is compelled to make up a game, a scenario....And the other thing is they have unwarranted faith in the accuracy of computers. Like I'd say, that just doesn't look right. Forget the learning package, it doesn't look right. Oh, well, you see, [they say] this is what fog is like, the computer says this is it. Come outside, [I say] that's fog.[35]

In an age of spectacular movie effects, minimal dialogue and media 'product' that can be sold globally because it does not depend on language, Curry staunchly defended the importance of stories and again, the need for realism and credibility drives his argument. He also reveals the process whereby experienced production workers pass on their expertise (or maybe not) to the younger generation of designers:

> Visual effects [can be] used as a crutch for poorly imagined stories ... we on *Star Trek* try to make the visual effects always drive the story, and we also try to avoid doing gratuitous ego gratification shots for the visual effects team. Because there are no limitations to what the computer can do, then the ability to do ridiculous, tasteless or unconvincing shots is there; so I try to get everybody to think that, OK, if we had a great cinematographer such as Haskell Wexler, and this was a real event, how would they photograph it? They would not photograph it with trombone turns, they would not do the zoom at a half mile and spin around the ship and then watch it go away. And so, by keeping the moves stately and constrained and supportive of the story, you never snap the audience's suspension of disbelief into a different reality by saying, 'Well, isn't this a cool shot that we did.' We're always trying to make sure that it never supersedes or snaps you away from the story.[36]

Internal evidence of the show's history

As Curry points out, design elements not only attest to changing tastes and technologies in the world at large, but also have their own internal histories as part of the history of the medium generally, and the series in particular: 'instead of just mimicking reality, you can create realities that are convincing, that can only exist in the medium, be it painting or be it movies, [these realities] do it better than any other medium so far,' said Curry. Herman Zimmerman, responsible for the overall design of the franchise, both television and movies and their spinoffs, had to be aware of continuity between the original series and *Enterprise* (whose story is set prior to the original series, with Kirk, Spock and co.). In doing so, he found that the design solutions produced by the original team in the 1960s, still worked in the 2000s:

> We used a lot of things that Gene [Roddenberry, the originator of *Star Trek*] invented. We're still using a flip communicator on the new series, we're still using the earpiece that Uhura [the communications officer, played significantly for the time, and even for these days, by a black actress, Nichelle Nichols] used in the classic series. We've even given Spock's viewfinder to T'Pol [the Vulcan female second in command in *Enterprise*].... If you notice it's a circular bridge. Voyager was not a circular bridge. We rather got away from the classic design of the bridge and so we've come back to that.[37]

Inevitably, in the more technologically sophisticated 2000s, there have been, as Zimmerman put it, 'some other wrinkles added to it'. They are using plasma screens, 'which are only inches thick instead of big boxy things that have to be hidden behind the walls'. For the designers, this creates internal verisimilitude and also connects more effectively with the contemporary audience:

> In a way we look less slick and that's part of the charm of the new series. You're closer to our own reality and the things you see on that bridge are – and in engineering and in the cabins, in the corridors. You'll find things that are more familiar to you. Not so imaginary. Not so far in the future.[38]

Jim Mees, the set-dresser, gave further testimony to the design authenticity, in order to create both a sense of being prior to the original series, but at the same time prefiguring it:

> What we've done on Enterprise now, is that everything is mechanical. It turns, it blinks. They have view finders, at Mayweather's station, Anthony who plays the character [the helmsman] sits at the helm. We actually went and bought plane parts so that there is a moveable steering wheel, there is a moveable gear shift.[39]

In those cases when the visuals of television as a source of information can be supported by inside information from the people who design and produce them, television scholars should attempt to validate their hypotheses about the history or influence of television by looking at aspects of agency and authorship among industry workers. Many assumptions about television's messages, techniques and effects may be altered when tested against empirical information about production processes. Production research is now a well-established part of research on television and has its own journal, the *Journal of Media Practice*. Ethnographies of production have also been carried out by John Tulloch and Albert Moran, and formal technical analyses, incorporating production studies, by John Caldwell.[40] In using television visuals as a source of evidence for the ways in which human beings behave, inferences made from the text itself can be enhanced, or challenged, by evidence from the human beings most directly involved in the relationship between that particular text and its 'readers': these are the people who produced the texts, and their audiences.

Audiences and fans

As mentioned, much of the evidence for television's impact actually comes not from an analysis of the texts themselves, and how they are constructed (including producer evidence) but from research with audiences. There has also been a great deal of research with children on the effects of television, both in terms of social learning and behavioural impact and in terms of cognition.[41] My own work on children and television drama, carried out on behalf of the BBC, and on children's understanding of the difference between fantasy and reality focused on children's awareness of drama as a constructed form: did they understand – as the Thermians in *Galaxy Quest* did not – that drama was a story, and how did they arrive at that understanding? What were the cues in what they were seeing (and hearing) that helped them make judgements about the events' reality and/or authenticity? The research took as axiomatic that the relationship between the viewers and the narrative was not necessarily a malevolent one: it consisted of ebbs and flows of understanding and emotion, and these perceptions and feelings were intrinsically related to the visuals and the sounds of what they were seeing.

In *Galaxy Quest* the role of the fans of the show is crucial; three groups of fans are represented. First, people who would be described disparagingly by *Star Trek* critics as 'Trekkies', that is people so devoted to the show that they dress up in the characters' costumes and focus their whole social lives on being a fan. Second, there are the alien 'Thermians', who go even further and take the show as gospel truth, as a simulacrum of reality, because their cultural expectations are different from those of the inhabitants of earth. Third, there are the 'geeks', the kind of 'techie' viewer described in Tulloch and Jenkins' fan study who went on to become space researchers when they grew up. The geeks or techies apply their love of the show to real-world activities, such as computer programming and technical design, and it is their combination of enthusiasm and rationality which eventually saves the day.

Perhaps the relationship between the audience and its loved text was best summarized by Patrick Stewart, an English Shakespearian actor widely admired and praised for his performance as Jean-Luc Picard, the captain of the Enterprise in *Star Trek: The Next Generation*:

> I've never understood what the benefit is in criticizing the fans. They after all made us a success and kept us on TV all those years and made us rich and famous. The majority of the fans are smart and interesting and enthusiastic and pay attention and respond well to what we do. . . . But whenever *Star Trek* fandom comes up people are only interested in the ones that dress as Klingons. . . . I know what it's like to stand in front of three thousand fans and the experience is a very, very positive experience. And that's why I won't join in the all too easy fun that's had at the expense of some of the extremists. Sometimes when you're working 12, 14, 16 hours a day you suddenly wonder who the hell am I doing this for. There were times when specific things would come up in some of the question and answer stuff [at fan conventions] that would be encouraging, and they were always so supportive about maintaining the overall seriousness of the show.[42]

Thus, Stewart questions Neil Postman's argument that it is not television's frivolity that is most harmful to society, but rather its attempts to tackle serious content. Stewart's years of experience in both legitimate theatre and in one of the most popular and talked-about drama series ever shown on television gives his argument some authority – it is for us, as scholars and viewers, to decide ultimately where the weight of the evidence lies.

Notes

1 Richard Howells, *Visual Culture* (Cambridge: Polity, 2003), 197.
2 *Galaxy Quest*, 1999, director Dean Parisot, production company, Dreamworks.
3 Maria Tatar, *The Hard Facts of the Grimms' Fairytales* (Princeton, NJ: Princeton University Press, 2003), 39.
4 See Neil Postman, *Amusing Ourselves to Death* (New York: Viking, 1985); Sut Jhally and Justin Lewis, *Enlightened Racism: The Cosby Show, Audiences and the Myth of the American Dream* (Boulder, CO: Westview, 1992); Justin Lewis, *The Ideological Octopus: An Exploration of Television and its Audience* (New York: Routledge, 1991).
5 For further and fuller information, see John Caughie, *Television Drama: Realism, Modernism and British Culture* (Oxford: Oxford University Press, 2000); Lez Cooke, *British Television Drama: A History* (London: British Film Institute, 2003); Jon Bignell and Stephen Lacey, *Popular Television Drama: Critical Perspectives* (Manchester: Manchester University Press, 2005); Jane Feuer, *MTM 'Quality Television'* (London: British Film Institute, 1984); Janet McCabe and Kim Akass, eds., *Quality TV: Contemporary American Television and Beyond* (London: Tauris, 2007).
6 Máire Messenger Davies, *'Dear BBC': Children, Television Storytelling and the Public Sphere* (Cambridge: Cambridge University Press, 2001).
7 Bob Hodge and David Tripp, *Children and Television* (Cambridge: Polity, 1986); Máire Messenger Davies, *Fake, Fact and Fantasy: Children's Interpretation of Television Reality* (Mahwah, NJ: Lawrence Erlbaum, 1997); Davies, 'Reality and Fantasy in the Media: Can Children Tell the Difference and How Do We Know?', in *The International Handbook of Children, Media and Culture*, ed. Sonia Livingstone and Kirsten Drotner (London: Sage, 2008), 121–36.
8 George Gerbner and Larry Gross, 'Living with Television: The Violence Profile', *Journal of Communication* 26, no. 2 (1976): 172–99; George Gerbner et al., 'Living with Television: The Dynamics of the Cultivation Process', in *Perspectives on Media Effects*, ed. J. Bryant and D. Zillman (Hillsdale, NJ: Lawrence Erlbaum, 1986); George Gerbner et al., 'The Demonstration of Power: Violence Profile No. 10', *Journal of Communication* 29 (1979), 177–96; Katherine Miller, *Communications Theories: Perspectives, Processes, and Contexts* (New York: McGraw-Hill, 2005).
9 Gavriel Salomon, *Interaction of Media, Cognition and Learning* (San Francisco, CA: Jossey-Bass, 1994, 1981).
10 Raymond Williams, *Television, Technology and Cultural Form* (London: Fontana, 1974).
11 John Ellis, *Visible Fictions: Cinema, Television, Video* (London: Routledge & Kegan Paul, [1982] 2000).
12 George Millerson, *The Technique of Television Production* (Oxford: Focal, 1990).
13 Karen Lury, *Interpreting Television* (London: Hodder Arnold, 2005).

14 Patrick Barwise and Andrew Ehrenberg, *Television and its Audience* (London: Sage, 1996).

15 Cooke, *British Television Drama*.

16 Helen Wheatley, ed., *Re-Viewing Television History: Critical Issues in Television Historiography* (London: Tauris, 2007).

17 See these at www.ahrc.ac.uk/news/news_pr/2007/ahrc_bbc_knowledge_exchange_programme.asp and www.ucl.ac.uk/filmstudies/movingimagearchives.

18 David Oswell, 'Early Children's Broadcasting in Britain, 1922–1964: Programming for a Liberal Democracy', *Historical Journal of Film, Radio and Television* 18, no. 3 (1997): 375–93.

19 Ken Jones and Hannah Davies, 'Keeping it Real: *Grange Hill* and the Representation of "The Child's World" in Children's Television Drama', in *Small Screens: Television for Children*, ed. David Buckingham (London: Leicester University Press, 2005), 151.

20 Máire Messenger Davies, 'Salvaging Television's Past: A Discussion of the Fates of Two 1970s Classic Serials, *Clayhanger* and *The Secret Garden*', in *Re-Viewing Television History: Critical Issues in Television Historiography*, ed. Helen Wheatley (London: Tauris, 2007), 40–52.

21 Jhally and Lewis, *Enlightened Racism*.

22 Martin Barker and Julian Petley, eds., *Ill Effects: The Media Violence Debate* (London: Routledge, 1997).

23 Jay Goulding, *Empire, Aliens, and Conquest: A Critique of American Ideology in* Star Trek *and Other Science Fiction Adventures* (Toronto: Sisyphus, 1985); Taylor Harrison et al., eds., *Enterprise Zones: Critical Positions on* Star Trek (Boulder, CO: Westview, 1996).

24 Daniel Bernardi, *Star Trek and History: Race-ing Toward a White Future* (New Brunswick, NJ: Rutgers University Press, 1988); Michael C. Pounds, *Race in Space: Ethnicity in Star Trek and Star Trek: The Next Generation* (Lanham, MD: Scarecrow Press, 1999); Robin Roberts, *Sexual Generations: Star Trek TNG and Gender* (Urbana, IL: University of Illinois Press, 1999).

25 John Tulloch and Henry Jenkins, *Science Fiction Audiences: Watching Star Trek and Doctor Who* (London: Routledge, 1995).

26 Máire Messenger Davies and Roberta E. Pearson, 'The Little Program that Could: The Relationship Between NBC and *Star Trek*', in *NBC: America's Network*, ed. Michele Hilmes (Berkeley, CA: University of California Press, 2007), 209–23; Davies and Pearson, 'To Boldly Bestride the Narrow World like a Colossus: Shakespeare, *Star Trek* and the European TV Market', in *European Culture and the Media: Changing Media, Changing Europe*, ed. Ib Bondebjerg and Peter Golding (London: Intellect, 2004), 65–90; Pearson and Davies, 'You're Not Going to See that on TV: *Star Trek: The Next Generation* in Film and Television', in *Quality Popular Television*, ed. Mark Jancovich (London: British Film Institute, 2003),103–17; Davies, 'Quality and Creativity in TV: The Work of Television Storytellers', in *Quality TV: Contemporary American Television and Beyond*, ed. Janet McCabe and Kim Akass (London: Tauris, 2007), 171–84.

27 Interview with Herman Zimmerman, Hollywood, CA, January 2002.

28 Ibid.

29 Interview with Robert Blackman, Hollywood, CA, January 2002.

30 Ibid.

31 Interview with Jim Mees, Hollywood, CA, January 2002.

32 Interview with Dan Curry, Hollywood, CA, January 2002.

33 Sue Palmer, *Toxic Childhood: How the Modern World is Damaging our Children and What We Can Do about It* (London: Orion, 2006).

34 Interview with Dan Curry.
35 Ibid.
36 Ibid.
37 Interview with Herman Zimmerman.
38 Ibid.
39 Interview with Jim Mees.
40 John Tulloch and Albert Moran, *A Country Practice: 'Quality Soap'* (Paddington, NSW: Currency Press, 1986); John Caldwell, *Televisuality: Style, Crisis, and Authority in American Television* (Brunswick, NJ: Rutgers University Press, 1995).
41 Dafna Lemish, *Children and Television: A Global Perspective* (Oxford: Blackwell, 2007; Davies, *The Future of Children's Television: Academic Literature Review* (Ofcom: www.ofcom.org, accessed 3 October 2007); Sonia Livingstone and Kirsten Drotner, eds., *The International Handbook of Children, Media and Culture* (London: Sage, 2008).
42 Interview with Patrick Stewart, London 2000.

11 The privileged discourse: advertising as an interpretive key to the consumer culture

by Jacqueline Botterill and Stephen Kline

> When the historian of the Twentieth Century shall have finished his narrative, and comes searching for the sub-title which shall best express the spirit of the period, we think it not at all unlikely that he may select 'the Age of Advertising' for the purpose.[1]

How prescient this writer seems, in the light of historian T. Jackson Lears' essay 'From Salvation to Self-Realization: Advertising and the Therapeutic Roots of the Consumer Culture 1880–1930', which uses advertising to expose the 'kernel of truth' at the core of Virginia Woolf's ironic comment that 'On or about December 1910, human character changed.' Lears claims a 'fundamental cultural transformation' *was* taking shape as nineteenth-century's Protestant ethos celebrating 'perpetual work, compulsive saving, civic responsibility and a rigid morality of self-denial' was supplanted by consumerist values which sanctioned 'periodic leisure, compulsive spending, apolitical passivity and an apparently permissive morality of individual fulfilment'. Lears notes: 'To thrive and spread, a consumer culture required more than a national apparatus of marketing and distribution; it also needed a favourable moral climate.' His historical account elaborates on advertising's pivotal role in the articulation of the therapeutic ethos arising from the complex historical dialectic between Americans' experience of rapid change and the advertising industry's rethinking of persuasion as 'each continually reshaped and intensified the other'. The pace of urbanization and immigration, he argues, fostered a unique sense of social upheaval and unreality left by the erosion of traditional values. To fill the void in personal life advertisers discovered they could arouse consumer demand by 'associating products with imaginary states of well-being'. As a result, advertising illustration came to emphasize self-fulfilment and self-development as they began speaking 'to many of the same preoccupations addressed by liberal ministers, psychologists, and other therapeutic ideologues'. He concludes: 'Their motives and intentions were various, but the overall effect of their efforts was to create a new and secular basis for capitalist cultural hegemony.'[2] Lears' work anticipated the growing interest in using advertising as evidence in studies of cultural history.

In a 2005 work *Social Communication in Advertising* William Leiss and his co-authors link the expanding culture of consumerism to the progressive commercialization of media wherein merchandisers were given ever-greater licence and means to talk to consumers.[3] So too, advertisers' ways of talking to consumers continued to diversify throughout the twentieth century guided by a singular purpose: to sell products in the mass market. Their work traces the century-long process of marketing experimentation overseen by agency personnel who progressively expand their means of communication by absorbing typography from print; illustration from news magazines; photography from fashion magazines; storytelling from radio; and dramatic tempo from films. The successive waves of commercialized media saw the voice of marketers become ever more pervasive, distinctive and potent until the dynamic institutional bridge between the burgeoning industrial economy and the mediated popular culture forged advertising into the *privileged discourse* of our global mediated marketplace.

Advertising is privileged for three reasons. First, advertising personnel became pivotal cultural intermediaries who greatly influenced the field of merchandising and product design. The agency system operated like a well-resourced laboratory, which not only applied consumer research extensively but also invented and tested evolving strategies of mediated communication.[4] Second, because advertising funded the expansion of newspapers, magazines, radio, TV, cinema and the Internet, its pragmatism invested the whole domain of mediated communication with a promotional logic. Unlike the arts, advertisers' pragmatic selling purpose ensured that audience considerations underwrote the circulation of popular culture.[5] Third, the agencies' economic leverage within the media-saturated marketplace ensures that messages about goods were constantly embedded in every aspect of mediated communication from art films to the Internet. Commercialism ensured that the marketer's voice is heard everywhere in our daily rounds – of driving, watching, walking the city street listening, reading and surfing the web. As Douglas Kellner says, 'We are living in one of the most artificial visual and image-saturated cultures in human history which makes understanding the complex construction and multiple social functions of visual imagery more important than ever before.'[6]

Given the prominence of advertising in our consumer culture this chapter overviews the work of historians who have turned to advertising as an *interpretive key* to map the consumer culture. Although most historians looked at the frozen tableaux offered by magazine ads, we refer to some more recent content analyses of TV ads, which focus critics on the continuing biases in values and aspirations, gender and ethnic stereotyping, and lifestyles. We conclude that, when approached as strategic social communication, the carefully contrived visualization of social life articulated in advertising offers a uniquely relevant kind of evidence for commenting on and analysing what Andrew Wernick called 'promotional culture'.[7]

Rethinking commercial culture

During the first half of the twentieth century, most humanities scholars followed Frank R. Leavis in dismissing advertising as a trivial and inconsequential facet of business communication, worthy of analysis only for its corrosive impact on our commercial culture.[8] Unlike art and literature, advertising was contaminated by its pragmatic selling purpose. As a literary form, copywriting seemed too facile, the imagery too transparent, the music too bombastic and insipid, and the social scenes too stereotypical to merit thoughtful commentary. Others dismissed advertising as capitalist propaganda buttressing a false consciousness through needs manipulation. Yet as the wartime factories retooled for peacetime consumerism, a few scholars began to rethink the important role commercial media was playing in the social reformation taking place in post-war America, seeing advertising as an interpretive keyhole into our increasingly affluent way of life. In 1951, Marshall McLuhan's *The Mechanical Bride* helped launch media studies with a thoughtful literary analysis of advertising, which apprehended it as the 'folklore' of industrializing society.[9] The scenes depicted by advertisers were not trivial and void of meaning, but condensed narratives – more like poems than novels, haiku than epics and ditties than operas. And, like poetry, their meaning had to be extracted through careful reading of the social relationships and situations in which goods were being incorporated into the industrial way of life.

In a series of 59 short essays, McLuhan highlighted the profound changes taking place in the American material culture as technologies – appliances, cars and media – become domesticated during the late 1940s. Advertisings' scenes from daily life provide a codex for the underlying patterns of technological change emerging in post-war America: a radio ad reveals the audience structure of mass communication in the primal scene of a family at the kitchen table listening (but not conversing) or eating; a refrigerator ad anticipates the paradoxical empowerment of the suburban housewife awaiting her family's return in the automated isolation of the suburban home. McLuhan's fascination with popular culture inspired a new way of interpreting the visual representations of things and their uses. His masterwork *Understanding Media* proclaimed that 'the ads of our time are the richest and most faithful daily reflections that a society ever made of its entire range of activities'.[10]

As foretold in *Printer's Ink*, the writings of Vance Packard, John Kenneth Galbraith, Herbert Marcuse and Theodor Adorno display a deepening fascination with the commercial dynamic of mass society.[11] Calling advertising a 'magical system', Raymond Williams argued that marketers' penetration of mass media rivalled traditional forms of public communication.[12] As agencies perfected their communicative techniques of mass marketing – branding, targeting and design – cultural theorists zoomed in on TV because it brought advertising's unique representation of contemporary social life into the home.[13] Building on McLuhan's media theories, Jean Baudrillard proclaimed that the 'telematic power of television' amounts to nothing less than 'a fundamental mutation in the ecology of the human species'. 'Today', he wrote, 'we are everywhere surrounded by the remarkable conspicuousness of consumption and affluence', which invades and falsifies everything. Marketing, purchasing, sales are so woven into the fabric of our lifestyles that they 'constitute our language, a code with which our entire society speaks of and to itself'.[14]

Decoding advertising

The writings of a generation of media theorists have focused critical attention on the unique imagery through which our 'culture speaks to itself' in the parallel promotional universe depicted in advertising. In the early 1970s French cultural theorist Roland Barthes called for a semiotic method to broach the language of 'objects' found in advertising. In his book *Mythologies*, Barthes claims that, to make sense of advertising representations, the analyst must go beyond the verbatim statements about the products; our decoding of the material world of goods requires uncovering the complex cultural value system that infuses consumption of specific things with broader significance.[15] Products, he argued, are not merely objects with defined functions, but goods embedded in a cultural system or 'code': as symbols, goods are lodged in the customs, values, attitudes and tastes of the culture that gives meaning to their use and display. For example, a rose in an ad invokes a relationship of romantic love between a man and woman sitting at a table because in western cultures roses are 'signifiers' of passionate love. To buy a rose for someone is a way of expressing an abstract feeling about him or her because the 'signifier' (image of the rose) is meaningfully connected to the abstract 'signified' (passionate love). The task of decoding an ad implies the identification of the meaningful connection between 'images' and the underlying cultural reference system – in this case one in which roses have come to be associated with 'love'.

Since Barthes first pointed to advertising's place in the cultural system, linguists have sought to reveal why consumers accept its false and misleading ideas.[16] To 'crack the code' of advertising one had to uncover the ideology, the means by which marketing distorts reality. Judith Williamson's *Decoding Advertisements* takes on this issue by deconstructing advertising design.[17] Williamson uses a simple Chanel ad to illustrate her approach to interpreting the ideology of advertising. The ad presents the photographed face of a beautiful woman (the French fashion model and actress Catherine Deneuve, who embodies French chic, glamour, beauty and sophistication) juxtaposed against a picture of the product (an elegant bottle of Chanel No. 5). Williamson explains that both these images are symbols well established within our cultural 'referent system' which define qualities of French beauty. Although Chanel and Deneuve are symbols that pre-exist the ad, they have been selected and arranged carefully by the advertiser to strengthen the association based on compositional conventions of photographic layout. It is only through the visual relationship between these two symbols that the ad designers invite the reader to believe that Chanel No. 5 is like the actress – chic, feminine, sophisticated and elegant – and, implicitly, that by buying it the user too becomes meaningfully linked to this epitome of French femininity and flawless beauty.

Ideology, Williamson notes, is a theory about the transfer of meaning between ad designers who 'encode' specific meanings and readers who 'decode' images. For an ad to have 'significance', there must be two kinds of labour performed: the first is the design work of the advertiser who composes the images based on their knowledge of the broader cultural repertoire and conventions of storytelling; the second is the reader who also has to do some 'interpretive work' – to make an effort to grasp what the ad designer is saying to them. This transfer of value is possible only if both share the same underlying visual symbols and visual conventions that connect them. Any attempt at textual analysis therefore must go beyond recognizing the biased representations embodied in advertising's underlying value system. The analyst must also explain why viewers accept the ideas and values invoked in the ad by *decoding its design* to explain how the advertisers selection and arrangement of visual symbols and words can shape how readers will interpret them.

In his *Gender Advertisements* (1979), Erving Goffman observes that most contemporary ads present images of stereotypical consumers engaging in ritual acts of consumption. Goffman's sociological decoding of advertising attempts to view the social world represented in advertising as we would a drama.[18] If not entirely natural, advertisers' staging of these social interactions are recognizable and familiar to us because the underlying protocols of social performance are known to all. Our cultural knowledge of social interaction not only enables us to perform appropriately in specific situations – a first date, an encounter with the boss, a party – but also enables us to read and judge the 'staged' social performances of the 'other' in advertising. Advertisers do not create those codes of cultural conduct but they do draw upon our knowledge of social communication: to ensure that their 'displays' are 'read' in the correct way. Goffman maintains that decoding the imagery of advertising therefore requires knowledge of the iconography of ritual displays. As Goffman says, 'Advertisers conventionalize our conventions, stylize what is already a stylization, make frivolous use of what is already something considerably cut off from contextual controls. Their hype is hyper-ritualization.'[19]

Goffman demonstrates the value of dramaturgical analysis through his deconstruction of gender imagery in advertising. Gender, of course, is one of the most durable forms of codified behaviour in all societies, and every culture has accepted 'routine' forms for communicating gender identity. A culture's social norms indicate how men and women are supposed to look, act and relate to each other in a wide variety of social situations; the resultant ritualized behaviour anchors expectations, rewards and punishments and stabilizes social intercourse. Ads have to communicate at a quick glance, and they require the participation of the audience to transfer their meaning. They therefore use 'body language'. The hug, smile and downcast gaze are part of the gestural code of masculinity and femininity that we learn from birth. For example, women's hands are usually portrayed just caressing or barely touching an object, as though they were not in full control of it, whereas men's hands are shown strongly grasping and manipulating objects. Female body language is one of abasement. Not only do women kneeling and lying down signal subservience, but also lying down is a 'conventionalized expression of sexual availability'. Women, he shows, are also infantilized, gazing wide-eyed at the camera or sucking their thumbs in nervous anticipation of their date's arrival at the door. What better source to draw upon than an area of social behaviour in which ritual gestures are instantly recognizable, and which touches the very core of our definition as human beings? Goffman's sociological analysis reveals that such repeated portrayals are not random or neutral, but biases in the relative social positions of men and women. In this sense, the cultural codes and conventions of design are what make the biased representations of advertising an accomplice in reinforcing submissiveness associated with femininity.

Linguists took up the notion of advertising as a language to further develop this ideological critique. In *The Language of Advertising* (1985), Torben Vestergaard and Kim Schroder focused on the hyperbolic images and exaggerated claims that advertisers made with impunity. The art of advertising is puffery: a nice name for a style of language use, which stretches the truth. In their mind, advertisers' ability to visually distort and narratively enhance consumers' everyday experiences subverts rational consumption choices.[20] Trying to explain advertising's persuasive impact, Paul Messaris applies Pierce's theory of iconic representation to explore the falsified verisimilitude of advertising images. Messaris claims that visual communication – whether it be photographs, drawings, films, TV or paintings – lacks an 'explicit syntax for expressing analogies, contrasts, causal claims, and other kinds of propositions'.[21] Yet upon studying a selection of magazine advertisements, he notes, ironically, that this ambiguity of iconic representations is an asset for advertisers.

Messaris believes that 'this relative indeterminacy of visual syntax plays a central part in the processes of visual persuasion' since the claims made visually in advertising can never have a precise logic. 'Iconicity', he claims, 'gives advertisers access to a broad spectrum of emotional responses that can be enlisted in the service of an ad's cause.'[22] Messaris concludes therefore that advertising's ideological impact is related to its visual rather than verbal claims: 'By drawing on their intuitive understanding as well as a growing body of research concerning the relationship between vision and emotion, advertisers are able to elicit strong, sometimes primal reactions.'

In *Undressing the Ad* (1998), Katherine Frith argues that decoding the values, social relations and ideology embedded in promotional storytelling is now a vital field in visual cultural studies and a valuable skill for those undertaking historical research.[23] She claims that 'the most useful technique for critically deconstructing both the surface and deeper social and cultural meaning of advertisements is a form of textual analysis'.[24] In their study of print advertisements, Rodney Heiligmann and Vickie Rutledge Shields explore how contemporary advertisers bring together stylized text and images that draw upon the symbolic repertoire of media literate audiences. Their work urges researchers to pay attention to text, image and audience to make sense of advertising.[25] As Chiara Giaccardi has pointed out, the task of interpreting the dramatic social scenes of TV ads requires care. Because it is neither an accurate mirror of society, nor a total simulacrum, making sense of advertising's 'commercial realism' requires understanding of its unique relationship to social reality:

> Rather than mirroring social reality, advertisements put it on stage, construct a discourse out of particular aspects, draw on topical issues and discursive conventions; they select from among a range of possibilities, related to both form and content, and elaborate a version of social reality which is neither 'true' nor 'false', and which often lacks verisimilitude, but is always meaningful.[26]

Deconstructing the appeal of advertising

A thoughtful reading of the social imagery of advertising is a valuable starting point for critiques of contemporary culture. But approaching advertising as a language or ideological text, rather than a unique design practice, has its limitations. Business historian Roland Marchand provides a historical explanation of advertisings' refocusing on social and psychological consumer experiences during the 1920s, by documenting how ad agencies became advocates of a novel conception of marketing that privileged consumer needs rather than product utility.[27] For example, innovation in consumer research in the marketing department at General Motors (GM) helped the automaker to design and sell cars on the basis of the consumer's preferences for styles of grille and colour rather than the Fordist appeal to price and utility. According to Marchand, GM's acknowledgement of the consumer, however minor, helped reposition GM as a democratic, fashionable and modern company. GM won half of the automobile market away from Ford during the 1930s based on this appeal. With ever greater numbers of similar consumer products flowing off the factory assembly-lines, merchandisers generally realized that their brands had to be designed and marketed with consumer 'desires and anxieties' in mind. The safety razor, agencies argued, could no longer be sold on the basis of need for sharpness or durability; first, the male populace had to be persuaded that the clean-shaven face was fashionable. The laundry detergent required medical experts inspecting the bathroom to convince women that their sinks were free of germs. Marchand claims that armed with the broader palette of expressive formats, marketers not only sold goods, but also legitimized a modern way of life that deployed packaged goods into solutions for the complex contradictions of daily life. Agencies took on a broader social role as the 'apostles of a modernity' – the leading advocates of a marketing method that focused on customers rather than engineering.

This reorientation to consumers had profound consequences for the way advertisers communicated about consumption. No longer simply a matter of demonstrating the products' benefits (reason-why formats), the task of copywriters required them to compose elaborate social narratives that helped reconcile the consuming public to the anxieties and complexities of twentieth-century life. As Marchand observes:

> The result of this trend toward emphasis on consumer satisfactions was called 'dramatic realism' – a style derived from the romantic novel and soon institutionalized in the radio soap opera. It intensified everyday problems and triumphs by tearing them out of humdrum routine, spotlighting them as crucial to immediate life decisions, or fantasizing them within enhanced, luxurious social settings. [28]

Guided by their new focus on consumer desire rather than product qualities, they innovated in storytelling and layout by adapting the compositional tropes and writerly contrivances from magazines, comics, films and radio. The whole gamut of popular culture was pillaged in the distillation of this distinctive rhetoric of selling to consumers. Focusing imaginatively on the motivations behind purchase, agencies took the lead in recognizing that women were primary consumers. Their copywriting appreciated women's changing roles and aspirations within the middle-class household. In ads, pre-baked bread and canned soup were not just practical solutions but moral dilemmas: 'Some people say that modern mothers are not such good home-makers as were the mothers of olden days,' the copywriters admitted. 'But that's not true,' they assured their gentle readers. 'Human nature doesn't change a great deal' and canned soup is just as nutritious. Such easily repeatable formulae about women's position as modern housewives helped bridge traditional ideas of patriarchy with the emerging normative order of a mass-consumer age in which 'ready made' and 'convenient' were pragmatic resolutions of their wifely duties. His analysis of appeals indicates how advertising not only introduced consumers to new products, but also transformed the world of goods into a therapeutic force by offering 'balms' to assuage the 'anxieties' that individuals experienced as they acclimatized to modernity.

Reviewing the design, layout and themes of 18,000 ads from the 1920s and 1930s, Marchand provides a magisterial overview of the changing marketing practices behind advertisers' storytelling. Rather than decoding an exemplary ad, he looks across the whole corpus to identify the narrative tropes and visual clichés, which underpin

advertisers' appeals to consumers. Marchand's historical method interprets the underlying motives and psychosocial sentiments associated with the tableaux and parables
found in advertising: the family circle, the hopeful view from the factory, the prideful
homeowner all give expression to the changes taking place in marketers' ways of thinking
about what motivates consumption. Marchand asks whether advertising's mirror 'provides benign, therapeutic deceptions rather than reflections of social reality'. His answer,
however, is complex: it depends upon what we look at and how we analyse it:

> If we focus on the cast of characters in the tableaux we might be impressed by
> the 'ads distortion of social circumstances' ... but if we focus on the perception of
> social and cultural dilemmas revealed in the tableaux, we will discover accurate,
> expressive images of underlying 'realities' ... reflected in advertising's elusive
> mirror.[29]

Form as content: the stylization of life

Cultural historian Stuart Ewen has similarly noticed the way design, packaging and
display came to dominate American merchandising during the 1920s.[30] The Calkins and
Holden agency were the forerunners of a new kind of marketing based on the adaptation
of aesthetics into the products, packaging and advertising. They even opened the Fashion
Coordination Bureau, which supplied 12,000 department stores with directives on how
to dress windows and decorate stores with aesthetics in mind. For Ewen, the importance
of advertisers' growing fascination with design was that marketers implicitly prioritize
style and form over function and utility. As image management became the central
strategy of commercial communications after the Second World War, 'the power of
provocative surfaces spoke to the eye's mind, overshadowing matters of quality or substance'. 'Style, more and more, has become the official idiom of the marketplace', making
the cultural references projected into goods the 'most constantly available lexicon from
which many of us draw the visual grammar of our lives. It is a behavioural model that is
closely interwoven with the modern patterns of survival and desire.'

Acknowledging the growing psychological sophistication of advertising's artistic
style and formats between 1910 and 1970, in *Fables of Abundance* T. Jackson Lears claims
that the novel visual rhetoric of 'advertisements has acquired a powerful iconic significance'. Ads have coupled words and pictures in commercial fables – 'stories that have
been both fabulous and didactic, that have evoked fantasies and pointed morals, that
have reconfigured ancient dreams of abundance to fit the modern world of goods'. Yet as
a historian of advertising design, Lears rejects critical theorists who dismiss advertising
solely as a manipulation of unwary consumers pointing instead to what he calls the
'unintended consequences' of advertisers' efforts to vend their wares garbed in art and
poetry: 'The creation of a symbolic universe where certain cultural values were sanctioned and others rendered marginal or invisible.'[31]

Lears' nuanced historical study reveals how the valorization of consumerism
emerged from a collaboration between advertisers and 'other institutions in promoting
what became the dominant aspirations, anxieties, even notions of personal identity in
the modern United States'. Reading the implicit normative order in these fables of

abundance, Lears detects a growing disenchantment with the managerial worldview, and the valorization of a romantic one in which individualism and self-expression ultimately triumph over mass conformity through a revival of animistic thought. He concludes that 'modern advertising could be seen less as an agent of materialism than as one of the cultural forces working to disconnect human beings from the material world'.[32]

John Berger's film and book *Ways of Seeing* brought an art historian's critical eye to the task of revealing the cultural values embedded in advertising narratives.[33] Berger noted that, however typical advertising's thematic preoccupations, it represented the material world visually, not in the form of an engineer's schematic, but by deploying the imaginative visual conventions borrowed from painting, film and fashion – whether it be the possessive male gaze at the perfected beauty of the female nude, the photographic tangibility of abundance in the oil painted still life, or the prideful domesticity of bourgeois portraiture. Advertising employed various artists to ensure the look of their imagery: fashion photographers to pose the models, set designers to chose the decor or art directors to ensure that the light glinted off the bottle in the product 'beauty shot'.

Berger shows how, in the camera's gaze, however mundane the objects, the possession and consumption of them became aestheticized through the skilful deployment of graphic conventions borrowed from the various visual arts, especially fashion photography and oil painting. The 'ways of seeing' found in TV advertising exemplified the underlying sensibilities of these borrowed techniques of display based on a dialogue of forms, a borrowing of design ideas from other forms of expression. Advertisers used the lexicon of design and the aura of the art object to imbue products with emotion beyond the mundane and prosaic. For example, using the technique of visual morphing, the Keri ad depicts the individual models as if they were emerging from well-known paintings. Their beauty achieved through the use of the product is thus fused with the beauty of these famous paintings, and indeed the very idea of art (Fig. 11.1). By using these conventions, advertising turned the material world of consumption into dreamscape where our lives gain depth and lustre through the visualization of a good's imaginary powers. Originally associated with painting, music, literature and photography, the aestheticization of lifestyle became a hard to define but easy to recognize aspect of the consumerist ideology. Given marketers' attention to aesthetics and style, historians of fashion find advertising to be an excellent corpus wherein to find a record of the ideas and values embedded in the changing design of goods.[34]

The values of advertising in a global market

Many cultural researchers have turned to content analysis to structure their temporal comparisons of advertising design. Richard Pollay's content analysis of the changing values reflected in a large sample of magazine ads covering the period from 1900 to 1979 finds that, rather than a moral and aesthetic revolution, advertising images reveal a slowly changing cluster of dominant values. Pollay notes that 'there are clearly some large-scale historical trends in advertising copy, in particular the trend toward selling consumer benefits, rather than product attributes and the trend toward creating favourable attitudes rather than communicating cognitive content'.[35] Pollay's study uses a typology of 42 values to gauge American perceptions of the world of goods more

With Keri enriched with Emmoliants and
vitamin E

Keri makes skin so touchable you'll feel like a woman
inside and out

Keri Forever Beautiful

Figure 11.1 Keri advertisement storyboard

precisely. Pragmatism and efficiency remain the dominant appeal throughout the century. Americans seem to want their cars and cough medicines to perform as promised as well as to be brightly coloured. Overall, Pollay notes that the values depicted in ads depend on product sector: cars and floor cleaners are sold in different ways from food and holidays. Luxury, beauty and glamour persist as values, especially in the fashion and personal care markets, while convenience and pride in one's house are increasingly repeated ideals mentioned in these mass marketing domestic slices of life after the war. Care for family and home, fun, pleasure and romance are prized too, but less so depending on the product sector. In conclusion, Pollay suggests that advertising's mirror reflects a slightly distorted image of the American way of life, one that highlights not who we are, but what might make us buy.

Increasingly, researchers wondered whether the globalization of commercial media resulted in differentiation or convergence of cultural values around the world.[36] They found that the fundamental transformations to consumerism that Lears identified were not restricted to the United States. A study by Robert Goldman and others showed the persistence of modernist ideals in the values communicated in 800 corporate television ads from the high-tech sector sampled globally: technology's triumph over space and time, the celebration of globalization and a reverence for individualism.[37] Modernist values of progress, science and optimism become enduring and universal values within these postmodern messages where individual style is fluid. In their view, advertising creates the foundation for global capitalism by painting a values landscape where technology provides the fix for the world's personal, social and political malaise.

However, other studies have pointed out that because some modes of consumption are deeply culturally embedded, western cultural values have not been fully adopted in Taiwanese advertising.[38] Kara Chan also finds that although technology ads are very similar in China and the United States, the 'correlation between product categories and cultural values is society based'.[39] Her study found differences in the food and drink product category advertised in China and Hong Kong. Chinese food and drink commercials mainly employed 'tradition', 'collectivism' and 'family' values, while Hong Kong food and drink commercials mainly displayed 'enjoyment', 'quality' and 'health' values. The results of global comparisons of advertising indicate that although global campaigns with uniform brand values can account for up to 30 per cent of advertising values – especially the hypermodernism associated with the global marketing of appliances, cars and media technologies – many other globally available products are advertised within cultural specific value systems. Cheng and Schweitzer identified eight cultural values dominant in either Chinese or US television commercials and reported that Chinese commercials resorted more often to symbolic cultural values while US commercials tended to use both symbolic and utilitarian ones.[40]

The privileged discourse and images of well-being

Leiss and co-authors set out to trace the implications of the agencies blending of design and consumer research into a new promotional rhetoric of social communication. Their historical study analyses 1800 ads drawn systematically from two magazines between 1920 and 1980 in an attempt to describe the subtle changes in not only the values but the appeals and images of well-being offered by the marketplace. Their study identifies five 'cultural frames', which exemplify the evolution of advertisers' strategies for branding, targeting and influencing consumers which add to the repertoire of appeals that can be used in advertising' attempts to promote goods:

- 1900–25 *Idolatry*: emphasis on material products and their benefits
- 1920–45 *Iconology*: emphasis on goods as symbols of modernity
- 1945–65 *Narcissism*: emphasis on consumption as expressive of an individual personality
- 1965–85 *Totemism*: emphasis on brands as badges of lifestyle communities
- 1980–2005 *Mise-en-scène*: articulating consumerism generally as a way of life that embraces the diversity of contemporary modes of consumption.

Advertisers were also found to exhibit a growing reliance on iconic codes of representation over lexical ones, finding that the visual register allows them to express the widening vista of consumer desires and anxieties. Guided by the marketing concept, advertisers broadened the formats of storytelling by including images, drawings, charts, text, typography, cartoons, slogans – and in TV, graphics, voice, music and sound effects – in the name of strengthening consumers' affective engagement with brand narratives. Images of people, objects and settings historically account for an ever greater proportion of the page layout, especially after colour, photography and illustration are introduced in the 1920s.[41] Yet, unlike fashion photography and art, advertising's persuasive rhetoric stresses the connection between images and words, relying on slogans, brand names, dialogue and explanatory text to strengthen the resonance of their brand stories with the experiences and lifestyles of specific market segments. Commercial realism speaks in psychological truths.

Adapting its formats and psychosocial appeals to successive commercial media, and guided by evaluative research, they conclude that advertising designers have acquired a deep yet pragmatic appreciation of the complexities of social influence in a cluttered mediated marketplace. Yet they insist that advertising genre cannot be understood as a unified language or read as homogeneous ideology. Advertising's stories about goods are flexible and layered: articulated around audiences, modes of consumption and market segments. Although the utility and functionality of goods feature prominently in the early ads, it is the objects' fashionability, social acceptability and individuality that come to the fore after the Second World War. Especially in the up-market magazines like *Vanity Fair*, the iconography of fluid taste cultures, aestheticization and celebrity lifestyles which seems to have replaced the vistas of status, luxury and usefulness in these images of well-being.

The growing emphasis on psychological states and social well-being is also confirmed by James McNeal and Stephen McDaniel's content analysis of television ads which examined 1682 need appeals in television advertising using Maslow's hierarchy of needs to map consumer motivations. Although food and soft drinks marketers often appeal to physiological need states (thirst quenching) and sensual experiences (lively taste), the vast majority of products in their sample embed consumption in the psychosocial values of social affiliation, self-esteem and avoiding embarrassment, autonomy and self-actualization. For example, appeals to esteem (36 per cent), self-actualization (25 per cent) and safety (12.6 per cent) predominate in car ads. They conclude that there is 'an evolution in need expression' taking place in the 1980s from a 'me' to a 'we' orientation in which the primary needs of thirst, nutrition and warmth were secondary to the psychological fulfilment and well-being sought through belonging, loving and sharing a way of being with others.[42]

In *Unconscious for Sale*, psychoanalysts Doris-Louise Haineault and Jean-Yves Roy examine 25,000 advertising posters between 1950 and 1990, reading the logic of 'desire' and 'fantasy' embedded in the compositional rhetoric of poster design. They explore the deeper psychological issues of individualism, power and sexualization that underpin the 'plastic figuration' that characterizes advertising imagery.[43] Analysing the psychosocial aspect of advertising appeals was also the focus of a psychodynamics-inspired study of British television and magazine ads between 1950 and 2000 by Barry Richards, Iain MacRury and Jackie Botterill. They found that advertising offered a rich array of symbolic resources designed to provoke unconscious fantasies, anxieties and archetypes for mental maturation processes. This study confirmed Pollay's point, finding that advertisers had not given themselves over to a celebration of hedonism, but rather to its tempered control. Refusing the psychoanalytic arguments offered by such authors as Christopher Lasch, who damned consumerism for fostering narcissism, Richards and his colleagues found a diverse array of 'object relations', which offered templates for psychosocial containment as much as it decontrolled individual desire and neurotic anxiety. What had shifted within the advertising of the second half of the twentieth century, however, were the waning appeals to social status and elitism. The authoritarian voice of the expert was replaced by softer more egalitarian invocation of the friend, and more individualized opportunities to find oneself – one's own identity – that acknowledged the diversity of consumers and the desire to customize one's lifestyle.[44]

The cast of characters: targeting, lifestyles and social representation

The implications of advertisers' growing articulation of new social roles and states of well-being is highlighted in Daniel Boorstin's *The Image* (1992), in which he argues: 'Advertising could not be understood as simply another form of salesmanship. It arrived at something new – the creation of consumption communities.'[45] In traditional societies, social categories are fairly fixed: usually defined by race, religion, gender and class. However, in a mass society, as occupational, race and gender distinctions were destabilized, consumption – one's lifestyle and preferences for brands – became more open and fluid. Advertising, Boorstin argued, contributed to this process through its narrow selection of the types of people and social settings it portrayed. Just as goods make 'visible and stable the categories of culture', the scenes in ads articulate the iconography of social belonging underlying daily life in our consumer society. By visualizing the stereotypical settings and lifestyles of various social categories, advertisers offered a vision of social life articulated around access to a constellation of goods. Thanks to advertising's permeation of popular culture, these new consumption communities begin to replace family, class and ethnic affiliations that once located people within the hierarchy of urban industrialism.

To deconstruct this social iconography, ad critics needed to explain how the typologies underwriting advertising's social communication – the patterned actions that structure social relations in specific settings or situations – are constructed. In *Social Communication in Advertising*, Leiss et al. argue that consumer orientation of mass marketing directed agencies to research human personality, motivations, social roles, life cycles, class, gender and ethnic divisions that pattern the purchase and use of goods. Based on their consumer research, marketers began to target their ads to specific consumer tastes and lifestyles. Over time their marketing segmentation became more refined and explicit. Sometimes called tribes and sometimes lifestyles, these marketer-driven consumer typologies became the basis of communication design, and were directed to specific market segments in media, which are prefigured around known audiences. The social constellations that unfold in ads are always carefully considered in relationship to the brand and the target market. It takes a lot of high-priced talent to ensure that the characters are dressed and groomed, the settings chosen and staged and the social interactions precisely photographed so that, from the hint of a smile to the tilt of the head, we can know whom the actors represent socially. To analyse advertising, therefore, requires attention to the lifestyle groupings and taste classes that are constantly forming and transforming in the mediated marketplace.

One of the first signs of advertisers' contribution to the articulation of consumer personae was the highly conventional representations of 'mass man' found in the 1950s. It was against this backdrop of middle-class banality and sameness, argues Thomas Frank, that advertisers themselves began to stress individual self-expression and standing out from others in a crowd, as themes in their ads. A subjectivity of stylized individuality associated with a taste for particular brands came to the fore during the 1960s, Frank says.[46] Amplified with computer graphics, director Zack Snyder explains on *Creative Club* how this idea inspired Miller Lite's visually stunning *Break from the Crowd* (Fig. 11.2). From the shopping and scrubbing suburban housewife of the 1950s to the metrosexual of the late 1990s, the categories of market segments have entered into our visualized persona of advertising: the YUPPY, Gen-X and the BoHo are but a few of the new consumption classes whose iconography cannot be understood outside of the consumer lifestyles and marketing research protocols which crystallize and dynamite them.

Frank Mort's *Cultures of Consumption* provides a detailed history of how the transition to a service economy in Britain through the 1980s led to the formation of new markets for fashion.[47] The growth of a service economy created a favourable environment for this new fusion of commodities and identity. The display of stylized commodities became central to not only social distinction but also the expression of self in a fluid landscape of social groups. The ideology of traditional masculine culture, which linked identity primarily to work, patterned male consumption around a limited set of goods and designated shopping as a female activity, was challenged within the new historical context. Advertisers, seeking to expand the circulation of commodities, hooked them into new versions of masculine identity. More than scripts of consumptions, the promotional messages invited contemplation of 'the interiority of experience' inviting young men to contemplate their narratives of self. Mort's work is significant in the detail it provides about how the new male personae were complexly woven within culture, through the work of not only advertisers but also fashion stylists, journalists, marketing executives, graphic designers and retailers. Together these cultural intermediaries fashioned novel

The concept for this commercial is a metaphor for breaking away from the crowd so the guys came up with this idea of using a big giant monster to represent the crowd and the creature is made up of people so he's a giant crowd walking through the city and our hero for miller lite destroys the creature

Figure 11.2 Miller Beer advertisement storyboard

masculine personas that hailed young men to think and recognize themselves through style. The penetration of the marketing effort into the domains of subjectivity raised complex questions about what it meant to be a man, and the new sexual politics that underwrote advertising's way of addressing this question.

Perhaps the most telling evidence of the importance of advertising's role in the articulation of social types, however, concerns the representation of African Americans.[48] For a long time ignored or stereotypically staged as servants and entertainers, African Americans were underrepresented in the ads of the first half of the twentieth century. Yet as they became acknowledged as part of the mass market, visible in popular culture as actors and musicians, the representation of black people increased, and the range of activities and goods they were associated with grew ever broader – from cosmetics to cars, advertisers started to hail blacks as consumers. In their study of the use of black actors in magazine and TV advertising between 1946 and 1986, George Zinkham, William Qualls and Abhijit Biswas note the dramatic embrace of blacks by the marketplace: from less than 1 per cent in the post-war period to 16 per cent in 1986 TV ads (approaching the actual share of the market).[49] A decade later, Robert Entman and Constance Book remarked that the inclusion of blacks into popular culture continued to increase, especially in TV advertising. No longer excluded or constrained to servile roles, about 30 per cent of their primetime sample included African Americans in various roles including parents, athletes, business people and social workers. Although they claim this indicates that 'blacks occupy a far less negative place in the symbolism of American culture than they once did', their content analysis of the staging of 'blackness' finds lingering stereotypes with more light- than dark-skinned models in ads.[50] Although dark skin is not associated with employment status or luxury goods, there is some indication that dark-skinned actors are used in mixed-race portrayals of social groups.

Gendered matter

The stereotypical representation of gendered relations, long accepted as normal, has been the most criticized aspect of commercial stereotyping. Clay McShane's *Down the Asphalt Path* supports his thoughtful cultural history of the automobile with a diagnosis of the promotional vistas of early-twentieth-century magazine advertising, which 'taught that cars allowed adventure and liberation from the everyday world for men through driving and for women through the prestige of being driven'. Even the ever practical Henry Ford, who bragged that his low-priced and functional products would sell themselves, was by 1907 advertising his Model Ns to women by associating the car with a 'romantic ride for two'. In women's magazines, rather than the roar of a powerful engine, a car purred; rather than the speed at which it traversed space, it was 'the gentle swaying motion of the car' that added allure.[51]

Especially since the 1970s, critics argued that advertising's perpetuation of gender stereotypes prevented equality between the sexes.[52] Pointing to marketers' contrived gender scenes, feminists claimed that marketers reinforced limited social status for women by showing them only as housewives, mothers and sex objects.[53] Studies of TV advertising provided ample evidence that advertising's restricted portrait of women's lives survived.[54] Feminists argued that only by deconstructing the assumptions behind the normalization of gender ideology could these attitudes change. Jean Kilbourne has noted, for example, how advertising's discourses amplified the problematic relationship between food and femininity. Kilbourne comments that

> food has long been advertised as a way for women both to demonstrate our love and to insure its requital. Countless television commercials feature a woman trying to get her husband and children to love her or just to pay attention to her via the cakes and breakfast cereals and muffins she serves.[55]

A recent ad for chocolate pudding mocks the secrecy surrounding women's intimate cravings by comparing sex and chocolate. But only the pleasure associated with eating chocolate is actually visualized, and in a manner that makes it seem erotic.

So, were advertisers merely reflecting existing gender identities or reinforcing inequalities by normalizing them through repetition? A few historical studies have attempted to evaluate whether the advertising industry has in fact adequately responded to the critiques of stereotyping offered by feminists during the 1980s and 1990s. Adrian Furnham and Elene Farragher undertook a comparative content analysis of 350 TV ads in the UK and New Zealand and concluded:

> The findings for the British sample of advertisements reveal that the advertising industry continues to portray men and women in decidedly different ways. The type of product and frequently the fictional context of the advertisement appear to dictate whether male or female actors or celebrities are used and how they behave. However, the degree of difference in the depiction of the sexes seems to have declined since the last investigations undertaken in Britain.[56]

However, they also found that the ads in New Zealand showed no signs of greater sensitivity to gender role representations.

Expanding the privileged discourse: marketing politics

To read advertising's persuasive imagery requires knowledge of the strategic protocols associated with branding, targeting and communicating with segmented markets. As Douglas Holt argues, brand management is a complex task. Brand's campaigns must communicate the qualities of an object or service by composing a brand story which conveys the qualities of desire and anxiety associated with a brand's relationship to targeted consumers. He sets out to describe the 'strategic language for brand management' by analysing successful television campaigns.[57] His approach is 'to analyze brands historically to uncover the principles that account for their success'.[58] According to Holt, brand stories become iconic when they offer consumers deeply meaningful myths that connect with the identity dilemmas thrown up by the complex shifting currents of consumer culture: iconic brands are associated with consumer identities not perceptions of product qualities. 'Iconic brands become the most compelling symbol of a set of ideas or values' which become the crystallization of a culturally situated desire or anxiety.

Today these brand strategies are executed in a multimedia commercial environment where access to segmented audiences is priced according to size and social composition. The most expensive cost more per second to make and show than a blockbuster movie. Because markets are competitive and media expensive, in all likelihood television campaigns are based on extensive market research about the consumers' perception of the competing brands in the category.

The promotional formats of advertising, originally developed to sell things, now serve as the pre-eminent communicative model to which all public interests have conformed. Education, healthcare and even dissenting advocacy groups like Greenpeace succumb to the logic of branding to promote their social and environmental ideals and projects. Commercial speech now rivals political speech as an ideological force in our public sphere. However, here we encounter one of the great paradoxes of marketing as both a critical reflection of and influence on culture. Advertisers' great facility at persuasive design makes them experts in both the perpetuation and deconstruction of social values. This social aspect of marketing has attracted the interest of mass culture critics recently as advertisers search for a way of making their promotional pitch less obtrusive and more socially responsible. Robert Goldman and Stephen Papson explain how experimentation with the ironic voice and self-mocking 'artificiality' are an important part of the emerging modes of address for a youthful generation who have become savvy about marketing communication.[59] Douglas Holt claims that advertisers have turned out to be consumerism's best ideological critics.[60] The airbrushed beauty of the cosmetic model satirized in Dove's *Campaign for Real Beauty* presents a deconstruction of the artificial surfaces of advertising's imagery far better than most feminist critics of the Beauty Myth (Fig. 11.3). For this reason, researchers have been particularly concerned by the promotional social influence applied by governments in social marketing, and political parties at elections, which utilize the visual salesmanship templates forged by the advertising industry.

The US literature points out three particular concerns about political marketing: the growing investment in television advertising as the fulcrum of political campaigning, the consequences of image politics (based on personality as opposed to policy issues being articulated) and the role of negative advertising and attack strategies in winning elections.[61] But as Hodess, Tedesco and Kaid have argued, this drift to negative advertising is not restricted to the United States. They cite growing concerns that UK 'party advertising has turned campaigns into more show than substance with these mediated bells and whistles leading democracy down the road toward the total packaging of politics'.[62] A content analysis of televised political messages that studied the type of spot and its issues, appeals and strategies, including non-verbal content and video production techniques, found that an increasing tone of UK political campaigning is becoming more personality focused and negative too.[63] Not only were the party election broadcasts more candidate driven but also the policies were made less explicit in the 1997 campaign than the 1992 one.

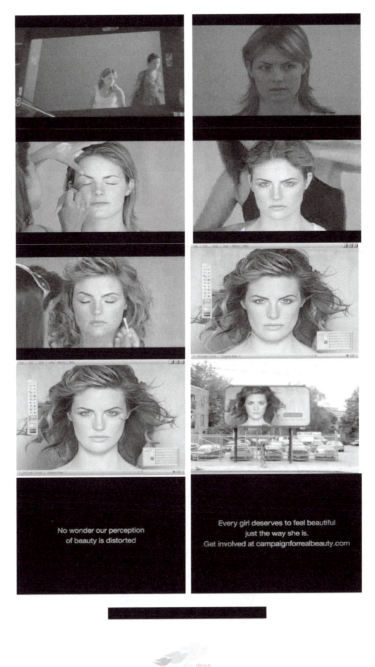

Figure 11.3 Dove advertisement storyboard

Figure 11.4 Sony advertisement storyboard

Conclusion

In a Sony ad we see advertisers crystallizing the emerging social rituals of the digital generation (Fig. 11.4). As a young man walks home down an ordinary urban streetscape, people everywhere not only look at him but also respond to him in a coy and overstated way. Schoolgirls flaunt their innocent sexuality, a model-beautiful woman pouts pro-vocatively from the hood of a car, an ageing labourer flexes his muscles and a couple with a baby smile inanely at the young man on the bus. Everywhere people turn to him and respond as if he emanates some magical force of attraction. On entering his flat, he finds his girlfriend already on the kitchen table, leaning back offering herself sensuously to his gaze. In a society where the sexualized gaze at strangers has been so carefully modulated, what can explain this persistent breach of social decorum? The answer is revealed in the final beauty-shot of the Sony camera: these are all people engaged in the familiar act of posing for the camera.

   This ad exemplifies state of the art television advertising design in several ways. First, it tells us little about the features of this digital technology, preferring instead to visually dramatize its magical powers over people. The ad illustrates the ideology of technological hyperbole so common in global marketing today. Second, its narrative (accompanied by a catchy back beat) takes the form of an engaging puzzle to the eye: it illustrates the ironic voice developed for a savvy youth market. Third, garbed in a familiar rock video aesthetic (perhaps sardonically referring to the opening scene in *Saturday Night Fever*), it also illustrates the dialogue of forms between advertising and popular culture. Fourth, the cast of characters is both specific and familiar: from sexy schoolgirls to big-smile Rastafarians. The stereotypes of commercial realism have a pristine clarity because they sit somewhere between our personal experiences of the street and its representation in film and tele-vision. Finally, the posed encounters are explicable only because in our culture, the camera as technology has helped to naturalize these para-social acts of ritual display – mugging, preening, pouting and aping for the camera – originally developed for the visual languages of fashion advertising and celebrity news. And here we confront the underlying point about social communication in the mediated marketplace: thanks to the camera and screen, the promotional tropes of marketing are familiar to us all. The ad reveals our hyper-modern subjectivity, trapped (as Andrew Wernick noted) in a promo-tional quagmire caught between a knowing scepticism of the ever-present promotional discourse and the psychic constancy of having to brand ourselves within the promotional culture through image management.

Notes

1   *Printer's Ink*, 1909.
2   Jackson Lears, 'From Salvation to Self-Realization: Advertising and the Therapeutic Roots of the Consumer Culture, 1880–1930', in *The Culture of Consumption: Critical Essays in American History, 1880–1980*, ed. R.W. Fox and T.J. Lears (New York: Pantheon, 1983), 4.
3   William Leiss et al., *Social Communication in Advertising: Consumption in the Mediated Marketplace* (New York: Routledge, 2005).

4 Linda Scott and Raheev Batra, *Persuasive Imagery: A Consumer Response Perspective* (Philadelphia, PA: Lawrence Erlbaum, 2003).

5 Jib Fowles, *Advertising and Popular Culture* (Thousand Oaks, CA: Sage, 1996).

6 Douglas Kellner, 'Critical Perspectives on Visual Imagery in Media and Cyberculture', *Cultural Studies Strikes Back*, www.gseis.ucla.edu/faculty/kellner/ed270/VISUALLIT critical.htm (accessed 4 May 2008).

7 Andrew Wernick, *Promotional Culture: Advertising, Ideology and Symbolic Expression* (Thousand Oaks, CA: Sage, 1991).

8 Frank R. Leavis, *Culture and Environment: The Training of Critical Awareness* (London: Chatto & Windus, 1933).

9 Marshall McLuhan, *The Mechanical Bride: Folklore of Industrial Man* (New York: Vanguard, 1951).

10 Marshall McLuhan, *Understanding Media: The Extensions of Man* (Boston, MA: MIT Press, 1964).

11 Vance Packard, *The Hidden Persuaders* (New York: Longmans, 1957); John Kenneth Galbraith, *The Affluent Society* (New York: Mentor, 1958); Herbert Marcuse, *One-Dimensional Man: Studies in the Ideology of Advanced Industrial Society*, 2nd ed. (Boston, MA: Beacon Press, [1964] 1991); Max Horkheimer and Theodor Adorno, *Dialect of Enlightenment* (New York: Continuum, [1944] 1987).

12 Raymond Williams, 'Advertising: The Magic System', in *Problems in Materialism and Culture* (London: Verso, 1980): 170–95.

13 Lawrence Samuel, *Postwar Television Advertising and the American Dream* (Austin, TX: University of Texas Press, 2001).

14 Jean Baudrillard, 'Consumer Society', in *Reflections on Commercial Life*, ed. Patrick Murray (New York: Routledge), 465.

15 Roland Barthes, *Mythologies*, trans. Annette Lavers (New York: Hill & Wang, 1984).

16 Bill Nichols, *Ideology and the Image: Social Representation in the Cinema and Other Media* (Bloomington, IN: University of Indiana Press, 1981); Stuart Ewen, *Captains of Consciousness: Advertising and the Social Roots of the Consumer Culture* (New York: McGraw-Hill, 1976).

17 Judith Williamson, *Decoding Advertisements: Ideology and Meaning in Advertisements* (London: Marion Boyars, 1978).

18 Erving Goffman, *Gender Advertisements* (Cambridge, MA: Harvard University Press, 1979), 23.

19 Ibid., 84.

20 Torben Vestergaard and Kim Schroder, *The Language of Advertising* (Oxford: Wiley Blackwell, 1985).

21 Paul Messaris, *Visual Persuasion: The Role of Images in Advertising* (Thousand Oaks, CA: Sage, 1997), xi.

22 Ibid.

23 Katherine Frith, *Undressing the Ad: Reading Culture in Advertising* (New York: Peter Lang, 1998).

24 Frith proposes the following general fields in reading the ad:
    1. read within the text (the surface meaning)
    2. retell the story (the intended meaning)
    3. identify the assumptions that privilege the text (the ideological meaning) (example of

the widened pupil signalling sexual arousal). (What makes the apparently natural ideological is the constructed meaning, i.e. the social relations.)

25 Rodney Heiligmann and Vickie Rutledge Shields, 'Media Literacy, Visual Syntax and Magazine Advertisements: Conceptualizing the Consumption of Reading by Media Literate Subjects', *Journal of Visual Literacy* 25 (2005): 41–66.

26 Chiara Giaccardi, 'A Comparative Study: Television Advertising and the Representation of Social Reality', *Theory Culture Society* 12 (1995): 113.

27 Roland Marchand, 'Customer Research as Public Relations: General Motors in the 1930s', in *Getting and Spending: American and European Consumer Society in the Twentieth Century*, ed. Susan Strasser et al. (Cambridge: Cambridge University Press, 1998).

28 Roland Marchand, *Advertising the American Dream* (Berkeley, CA: University of California Press, 1985), 24.

29 Ibid., 360.

30 Stuart Ewen, *All Consuming Images: The Politics of Style in Contemporary Culture* (New York: Basic Books, 1990).

31 Jackson Lears, *Fables of Abundance* (New York: Basic Books, 1994), 3.

32 Ibid., 4.

33 John Berger, *Ways of Seeing* (New York: Penguin, 1972).

34 Christopher Breward, Jenny Lister and David Gilbert, eds., *Swinging Sixties* (London: V&A Publications, 2006).

35 Richard Pollay, 'Twentieth Century Magazine Advertising: Determinants of Informativeness', *Written Communication* 1 (1984): 73.

36 Stephen Kline, 'Image Politics: Negative Advertising Strategies and the Election Audience', in *Buy this Book: Studies in Advertising and Consumption*, ed. Mica Nava et al. (London: Taylor & Francis, 1996), 139–56.

37 Robert Goldman, Stephen Papson and N. Kersey, *Landscapes of Global Capital – Commodification*, http://it.stlawu.edu/~global/tscg.bib.html (accessed April 2008).

38 James Tsao, 'Advertising and Cultural Values: A Content Analysis of Advertising in Taiwan', *International Communication Gazette* 53 (1994): 93–110.

39 Kara Chan, 'Cultural Values in Hong Kong Newspaper Advertising, 1946–1996', *International Journal of Advertising* 18 (1999): 437–554; see also N.D. Albers-Miller and B.D. Gelb, 'Business Advertising Appeals as a Mirror of Cultural Dimensions: A Study of Eleven Countries', *Journal of Advertising* 25 (1996): 57–71.

40 H. Cheng and J.C. Schweitzer, 'Cultural Values Reflected in China and US Advertising', *Journal of Advertising Research* 36 (1996): 27–41.

41 Leiss et al., *Social Communication in Advertising*.

42 James U. McNeal and Stephen W. McDaniel, 'An Analysis of Need-Appeals in Television Advertising', *Journal of the Academy of Marketing Science* 12 (1984): 187.

43 Doris-Louise Haineault and Jean-Yves Roy, *Unconscious for Sale: Advertising Psychoanalysis and the Public*, Theory and History of Literature 86 (Minneapolis, MN: University of Minnesota Press, 1993).

44 Barry Richards, Iain MacRury and Jackie Botterill, *The Dynamics of Advertising* (London: Routledge, 2000).

45 Daniel Boorstin, *The Image: A Guide to Pseudo-Events in America* (New York: Vintage, 1992), 145.

46 Thomas Frank, *The Conquest of Cool: Business Culture, Counterculture and the Rise of Hip Consumerism* (Chicago, IL: University of Chicago Press, 1997).

47 Frank Mort, *Cultures of Consumption: Masculinities and Social Space in Late Twentieth-Century Britain* (London: Routledge, 1996).

48 William O'Barr, *Culture and the Ad: Exploring Otherness in the World of Advertising* (Boulder, CO: Westview, 1994).

49 George Zinkhan, William Qualls and Abhijit Biswas, 'The Use of Blacks in Magazine and Television Advertising: 1946 to 1986', *Journalism Quarterly* 67 (1990): 547–53.

50 Robert Entman and Constance Book, 'Light Makes Right: Skin Colour and Racial Hierarchy in Television Advertising', in *Critical Studies in Media Commercialism*, ed. Robin Andersen and Lance Strate (Oxford: Oxford University Press, 2000), 214.

51 Clay McShane, *Down the Asphalt Path* (New York: Columbia University Press, 1995), 139.

52 Janice Winship, 'Sexuality for Sale', in *Culture, Media, Language*, ed. Stuart Hall (London: Hutchinson, 1980), 217–26; Jane Root, 'Who Does this Ad Think You Are?', in *Pictures of Women: Sexuality*, ed. Jane Hawksley (New York: HarperCollins, 1984), 51–68; Annette Kuhn, *The Power of the Image: Essays on Representation and Sexuality* (London: Routledge, 1985).

53 A. Courtney and T. Whipple, 'The Image of Women in Network TV Commercials', *Journal of Broadcasting* 17 (1974): 110–18.

54 Adrian Furnham and Elena Farragher, 'A Cross-Cultural Content Analysis of Sex-Role Stereotyping in Television Advertisements: A Comparison between Great Britain and New Zealand', *Journal of Broadcasting and Electronic Media* 44 (2000): 415–36.

55 Jean Kilbourne, *Can't Buy My Love: How Advertising Changes the Way We Think and Feel* (New York: Touchstone, 1999), 109.

56 Furnham and Farragher, 'A Cross-Cultural Content Analysis of Sex-Role Stereotyping in Television Advertisements', 434.

57 Douglas Holt, *How Brands Become Icons: The Principles of Cultural Branding* (Boston, MA: Harvard Business School Press, 2004), xii.

58 Ibid., xi.

59 Robert Goldman and Stephen Papson, *Sign Wars* (New York: Guilford, 1996).

60 Douglas Holt, 'Why Do Brands Cause Trouble? A Dialectical Theory of Consumer Culture and Branding', *Journal of Consumer Research* 29 (2002): 70–88.

61 Lynda Kaid and Anne Johnston, 'Negative versus Positive Television Advertising in US Presidential Campaigns, 1960–1988', *Journal of Communication* 41 (1991): 53–64.

62 Robin Hodess, John C. Tedesco and Lynda Lee Kaid, 'British Party Election Broadcasts: A Comparison of 1992 and 1997', *Harvard International Journal of Press/Politics* 5 (2000): 59.

63 Margaret Scammell and Holli Semetko, 'Political Advertising on Television: The British Experience', in *Political Advertising in Western Democracies*, ed. Lynda Kaid and Christina Holtz-Bacha (London: Sage, 1995).

Index